AF358622

Soft Photonic Crystals and Metamaterials

Soft Photonic Crystals and Metamaterials

Editors

Ivan V. Timofeev

Wei Lee

MDPI • Basel • Beijing • Wuhan • Barcelona • Belgrade • Manchester • Tokyo • Cluj • Tianjin

Editors
Ivan V. Timofeev
Laboratory for Photonics of
Molecular Systems
Kirensky Institute of Physics
FRC KSC SB RAS
Krasnoyarsk
Russia

Wei Lee
College of Photonics
Tainan Campus
National Yang Ming Chiao
Tung University
Tainan
Taiwan

Editorial Office
MDPI
St. Alban-Anlage 66
4052 Basel, Switzerland

This is a reprint of articles from the Special Issue published online in the open access journal *Materials* (ISSN 1996-1944) (available at: www.mdpi.com/journal/materials/special_issues/soft_photonic_crystals_metamaterials).

For citation purposes, cite each article independently as indicated on the article page online and as indicated below:

LastName, A.A.; LastName, B.B.; LastName, C.C. Article Title. *Journal Name* **Year**, *Volume Number*, Page Range.

ISBN 978-3-0365-6074-8 (Hbk)
ISBN 978-3-0365-6073-1 (PDF)

Contents

Preface to "Soft Photonic Crystals and Metamaterials" . . . . **vii**

Ivan V. Timofeev and Wei Lee
Special Issue: Soft Photonic Crystals and Metamaterials
Reprinted from: *Materials* **2022**, *15*, 8096, doi:10.3390/ma15228096 . . . . **1**

Heng-Yi Tseng, Li-Min Chang, Kuan-Wu Lin, Cheng-Chang Li, Wan-Hsuan Lin and Chun-Ta Wang et al.
Smart Window with Active-Passive Hybrid Control
Reprinted from: *Materials* **2020**, *13*, 4137, doi:10.3390/ma13184137 . . . . **5**

Junki Yokota, Kohsuke Matsumoto, Koji Usui, Shoichi Kubo and Atsushi Shishido
Effect of Host Structure on Optical Freedericksz Transition in Dye-Doped Liquid Crystals
Reprinted from: *Materials* **2022**, *15*, 4125, doi:10.3390/ma15124125 . . . . **15**

Toya Seki, Yutaro Seki, Naoto Iwata and Seiichi Furumi
Size-Controllable Synthesis of Monodisperse Magnetite Microparticles Leading to Magnetically Tunable Colloidal Crystals
Reprinted from: *Materials* **2022**, *15*, 4943, doi:10.3390/ma15144943 . . . . **27**

Silvia Adriana Estrada Alvarez, Isabella Guger, Jana Febbraro, Ayse Turak, Hong-Ru Lin and Yolanda Salinas et al.
Synthesis and Spatial Order Characterization of Controlled Silica Particle Sizes Organized as Photonic Crystals Arrays
Reprinted from: *Materials* **2022**, *15*, 5864, doi:10.3390/ma15175864 . . . . **39**

Inge Nys, Brecht Berteloot and Kristiaan Neyts
Photoaligned Liquid Crystal Devices with Switchable Hexagonal Diffraction Patterns
Reprinted from: *Materials* **2022**, *15*, 2453, doi:10.3390/ma15072453 . . . . **61**

Chia-Hua Yu, Po-Chang Wu and Wei Lee
Polymer Stabilization of Uniform Lying Helix Texture in a Bimesogen-Doped Cholesteric Liquid Crystal for Frequency-Modulated Electro-Optic Responses
Reprinted from: *Materials* **2022**, *15*, 771, doi:10.3390/ma15030771 . . . . **73**

Rashid G. Bikbaev, Dmitrii N. Maksimov, Kuo-Ping Chen and Ivan V. Timofeev
Double-Resolved Beam Steering by Metagrating-Based Tamm Plasmon Polariton
Reprinted from: *Materials* **2022**, *15*, 6014, doi:10.3390/ma15176014 . . . . **89**

Victor Yu. Reshetnyak, Igor P. Pinkevych, Timothy J. Bunning and Dean R. Evans
Influence of Rugate Filters on the Spectral Manifestation of Tamm Plasmon Polaritons
Reprinted from: *Materials* **2021**, *14*, 1282, doi:10.3390/ma14051282 . . . . **97**

Anastasia Yu. Avdeeva, Stepan Ya. Vetrov, Rashid G. Bikbaev, Maxim V. Pyatnov, Natalya V. Rudakova and Ivan V. Timofeev
Chiral Optical Tamm States at the Interface between a Dye-Doped Cholesteric Liquid Crystal and an Anisotropic Mirror
Reprinted from: *Materials* **2020**, *13*, 3255, doi:10.3390/ma13153255 . . . . **109**

Meng-Ying Lin, Wen-Hui Xu, Rashid G. Bikbaev, Jhen-Hong Yang, Chang-Ruei Li and Ivan V. Timofeev et al.
Chiral-Selective Tamm Plasmon Polaritons
Reprinted from: *Materials* **2021**, *14*, 2788, doi:10.3390/ma14112788 . . . . **119**

Preface to "Soft Photonic Crystals and Metamaterials"

This Special Issue on "Soft Photonic Crystals and Metamaterials" from *Materials* consists of 10 papers that highlight recent advances in a broad scope of optical-wavelength and sub-wavelength structures made of soft materials and particles. Soft matter shows plenty of unique and improved optical properties for deep scientific understanding, thereby promoting fabrication, characterization and device performance for potential photonic applications that include, but are not limited to, photovoltaic cells, photodetectors, light-emitting diodes, tunable microlasers, optical filters for biosensors, smart windows, virtual/augmented reality head-mounted elements, and high-speed spatial light modulators in glasses-free 3D displays.

Ivan V. Timofeev and Wei Lee

Editors

Editorial

Special Issue: Soft Photonic Crystals and Metamaterials

Ivan V. Timofeev [1,2,*] and Wei Lee [3,*]

1   Kirensky Institute of Physics, Federal Research Center KSC SB RAS, 660036 Krasnoyarsk, Russia
2   Siberian Federal University, 660041 Krasnoyarsk, Russia
3   Institute of Imaging and Biomedical Photonics, College of Photonics, National Yang Ming Chiao Tung University, Guiren District, Tainan 711010, Taiwan
*   Correspondence: tiv@iph.krasn.ru (I.V.T.); wei.lee@nycu.edu.tw (W.L.)

Citation: Timofeev, I.V.; Lee, W. Special Issue: Soft Photonic Crystals and Metamaterials. *Materials* **2022**, *15*, 8096. https://doi.org/10.3390/ma15228096

Received: 6 November 2022
Accepted: 13 November 2022
Published: 16 November 2022

Publisher's Note: MDPI stays neutral with regard to jurisdictional claims in published maps and institutional affiliations.

Soft matters include polymers, liquid crystals, colloids, biological tissues, and many smart materials. Weak molecular interactions make them sensitive to tiny mechanical, thermal, electric, optical, and other external stimuli. Photonic applications often require soft materials with significant or, in some cases, extreme optical dispersion and non-linear and anisotropic properties. The optical response can be structurally amplified in two fairly distinct scales that are wavelength-comparable periods of soft photonic crystals and subwavelength-graded soft photonic metamaterials.

The reader is invited to explore the findings of interdisciplinary materials science research included in this Special Issue, including fresh concepts from physics, chemistry, biology, and technology. This Special Issue presents ten original open-access articles, of which three are editor's choice articles. The collection can be divided into four categories. There are two papers related to smart windows (Tseng et al. [1] as well as Yokota et al. [2]). Two papers involve microparticle arrangement in colloidal and photonic crystalline structures (Seki et al. [3] and Alvarez et al. [4]). There are three related to switchable or tunable liquid crystal devices (Nys et al. [5], Yu et al. [6], and Bikbaev et al. [7]). The latter exploits light localization between the grating and the photonic crystal. Predicted by Tamm in 1932, this localized spot of light is called a Tamm plasmon, as it is a photon analogy of the electron state localized near the surface of a solid crystal. Finally, there are three more articles focusing on Tamm plasmons (Reshetnyak et al. [8], Avdeeva et al. [9], and Lin et al. [10]).

Tseng and coworkers [1] present a novel smart window utilizing salt and photochromic dichroic dye-doped cholesteric liquid crystal, that is essentially a soft photonic crystal. The smart window provides adjustable sunlight intensity and protects the privacy of people in a building. The functionality is based on the combination of two features: the darkening of a photosensitive layer upon exposure to ultraviolet light, and the response of cholesteric to voltage pulses. By independently switching both functionalities, four stable optical states are obtained: transparent, scattering, dark-clear, and dark-opaque. The presented work brings an interesting set of experimental data, including voltage, spectral, and temporal dependences, along with colored photos of the achieved optical states. The smart window offers passive automatic dimming and active haze control for privacy protection, exhibiting significant potential for applications.

The second paper on smart windows is contributed by Yokota et al. [2]. Fréedericksz transition in liquid crystals is one of the most classic and important concepts with real applications for liquid crystals in the society. Optical Fréedericksz transition is free of electrodes or wires and potentially more robust, however, it is not so popular because of the high threshold for light intensity. The paper strives to reduce the threshold until sunlight showing the dye doping has a great impact on the optical Fréedericksz transition, and that the host liquid crystal type also plays an important role. This is explained by the changes in the optical-electric torque, the elastic torque, and the dye torque. The threshold light intensity in trifluorinated liquid crystals is shown to be 42% lower than that in liquid crystals without fluorine substituents.

Colloidal and photonic crystalline structures are often fabricated by the arrangement of monodisperse microparticles. Seki and colleagues [3] report a one-pot hydrothermal synthesis of monodisperse magnetite microparticles. By adjusting the concentration of the reaction solution, the microparticle's diameter is manipulated from 100 to 200 nm. Under an external magnetic field the microparticles in aqueous suspension form a periodic structure that possess Bragg reflection colors observed in the spectral range from 730 nm to 570 nm by increasing the magnetic field.

The paper by Alvarez et al. comes from an international research group combining seven members from Austria, Canada, and Taiwan [4]. It presents opalescent photonic crystals obtained via self-assembly of silica microparticles using a solvent evaporation method. Silica microparticles are synthesized by the sol-gel method. The microparticle diameter is precisely controlled by only changing the amount of the solvent used. Thorough characterization of quasi-hexagonal spatial order is presented in Appendix using Voronoi tessellations, pair correlation functions, and bond order analysis. In both Seki et al. [3] and Alvarez et al. [4], the Bragg functionality of microparticle arrays is suggested for full-color reflective displays, sensing materials, solar cells, and coatings.

Switchable diffraction gratings and patterns are a promising class of soft-matter metasurfaces for modulated electro-optic devices. Nys et al. [5] use a thin liquid crystal layer with periodic photoalignment at the surfaces to fabricate stable two-dimensional hexagonal diffraction patterns. The research is brilliant both in experiment and simulation. First, the patterns between a strong scattering state and a near clear state on linear variation in the photoalignment direction are studied by means of polarization optical microscopy and light diffraction. Then, the numerically restored director configuration reveals a 3D superstructure, where all the twist conflicts in the confining substrates produce out-of-plane director reorientation without disclinations. The three-fold and six-fold twist angles yield hexagonal ordering that can be reversibly and continuously switched by small voltage.

Yu et al. [6] have investigated the uniform lying helix texture in a polymer-network cholesteric liquid crystal layer composed of a binary achiral mixture (of E7 and the bent-core dimer CB7CB). This 5 μm thick soft-matter film has fascinatedly tunable helicity with its helical axis along the substrate plane and its pitch of around 100 nm, which is shorter than the wavelengths of visible light. Applied voltage can generate two types of helix reorientation. At low voltage, the flexoelectric effect produces periodic splay-bend deformation of in-plain optical axis. Beyond the threshold voltage the dielectric effect occurs and the helix is gradually unwound until transformed to the helix-free homeotropic state. Polymer networks are used to stabilize this state against other orientations with lower free energy, such as planar, focal conic, and 3D-helical blue-phase states. A bifunctional monomer and a trifunctional monomer are utilized to demonstrate the advantage of photo-copolymerization for a temperature-stable cell. The final cell optical properties are guaranteed to be reversibly controlled by both voltage frequency and amplitude. A detail account of the novel incorporation of the liquid crystal dimer CB7CB is clearly stated in the paper.

Bikbaev et al. [7] are engaged in a challenge for mirrorless beam steering by generalized diffraction in voltage-tunable metagratings. With this purpose in mind, they consider a subwavelength grating patterned on top of a Bragg reflector. Under the metagrating strips the electromagnetic field can be efficiently localized in above-mentioned Tamm plasmon. This narrow resonance is extremely sensitive to refractive index of a top layer, which makes it possible to control each strip radiation phase in a wide range. This effect can be realized by the top layer made of transparent conductive material, such as indium tin oxide. The bias voltage up to 5 V leads to an increase in the volume concentration of the charge carriers in few nanometer vicinity of the layer boundary. As a result, the real part of the complex dielectric permittivity becomes negative, acquires metallic properties, and shifts the scattering phase by more than 200 degrees. Changing the number of strips with different applied bias voltage allows for efficient switching in first-order diffraction. In addition, it is shown that the reflected beam can be controlled by a continuous phase change along the metagrating. Definitely, this method increases the beam steering angular resolution.

Reshetnyak et al. [8] focus on the above-mentioned Tamm plasmons for the case of a thin silver layer adjacent to a rugate filter—a dielectric thin film with a smooth periodic profile of the refractive index, acting like a Bragg mirror. The smoothness helps to suppress sidelobes outside of a photonic band gap and provides several extra degrees of freedom for optimizing optical characteristics compared to multilayer Bragg mirrors. Surprisingly, the harmonic refractive index profile allows for convenient analytical expressions in regard to Tamm plasmon transmission and reflection peaks and dips. The analytics shows dependence on refractive index contrast, the metal layer thickness, and the external medium refractive index. As an example, it is shown that the Tamm plasmon wavelength can be at any point within the photonic band gap determined by the refractive index profile. The analytical approach complies quite well with exact numerical simulations in a broad range of experimentally available parameters.

Avdeeva et al. [9] investigate chiral Tamm plasmons—a class of the aforementioned Tamm plasmons with the broken mirror symmetry. The chirality of state originates from cholesteric liquid crystal, a soft self-organizing photonic crystal. A crucial requirement for such a state emergence is the presence of a polarization-preserving anisotropic mirror instead of conventional metallic layer. This mirror is not necessarily chiral, and therefore preserves the handedness of reflected light polarization compared to incoming one. Extensive dye-doping makes cholesteric highly dispersive material leading to spectral splitting of resonant features, such as the edge and defect modes of photonic crystal. The paper proves that chiral Tamm state spectral peak can be resonantly split as well, which is important for tunable chiral microlasers. The splitting is most pronounced when dye resonance frequency coincides with the frequency of the Tamm state. In this case, the reflectance, transmittance, and absorptance spectra show two distinct Tamm modes. For both modes, the field localization is at the interface of the polarization-preserving anisotropic mirror and the cholesteric.

Chiral Tamm plasmon was for a long time considered as a fantastic theoretical monster because of serious experimental restrictions imposed by the polarization-preserving anisotropic mirror. Lin et al. [10] proudly win the challenge of in-principle experimental realization of this chiral monster. They adopt a recently suggested 200 nm thick reflective half-wave phase plate implemented as a golden metasurface. Simultaneously they use an extremely birefringent stable cholesteric liquid crystal (Δn = 0.33). The 3 μm thick cholesteric layer provides almost optimal coupling of chiral Tamm plasmon when reflectance falls down to 20%. Moreover, the suggested design renders both phase and polarization matching. Furthermore, by temperature-changing the center wavelength of the cholesteric band gap with different pitches, the resonance wavelength of Tamm plasmons is tuned flexibly. This phenomenon is essential for advancing technical applications in novel types of optical switches, biosensors, polariton microlasers, and mirrorless lidars.

We would like to thank all the international teams of creative intellectuals and specialists who have made this Special Issue possible. First are the authors who have kindly prepared their contributed or invited papers. In addition, we would like to thank all the peer reviewers who participated in this Special Issue for their invaluable time to offer professional and constructive comments, which definitely helped promote the quality of the scientific content of each published paper. Our gratitude is extended to the people we interacted with at MDPI for their hard work and considerate coordination throughout the entire editorial process.

Funding: This research received no external funding.

Conflicts of Interest: The authors declare no conflict of interest.

References

1. Tseng, H.-Y.; Chang, L.-M.; Lin, K.-W.; Li, C.-C.; Lin, W.-H.; Wang, C.-T.; Lin, C.-W.; Liu, S.-H.; Lin, T.-H. Smart Window with Active-Passive Hybrid Control. *Materials* **2020**, *13*, 4137. [CrossRef] [PubMed]
2. Yokota, J.; Matsumoto, K.; Usui, K.; Kubo, S.; Shishido, A. Effect of Host Structure on Optical Freedericksz Transition in Dye-Doped Liquid Crystals. *Materials* **2022**, *15*, 4125. [CrossRef] [PubMed]
3. Seki, T.; Seki, Y.; Iwata, N.; Furumi, S. Size-Controllable Synthesis of Monodisperse Magnetite Microparticles Leading to Magnetically Tunable Colloidal Crystals. *Materials* **2022**, *15*, 4943. [CrossRef] [PubMed]
4. Estrada Alvarez, S.A.; Guger, I.; Febbraro, J.; Turak, A.; Lin, H.-R.; Salinas, Y.; Brüggemann, O. Synthesis and Spatial Order Characterization of Controlled Silica Particle Sizes Organized as Photonic Crystals Arrays. *Materials* **2022**, *15*, 5864. [CrossRef] [PubMed]
5. Nys, I.; Berteloot, B.; Neyts, K. Photoaligned Liquid Crystal Devices with Switchable Hexagonal Diffraction Patterns. *Materials* **2022**, *15*, 2453. [CrossRef] [PubMed]
6. Yu, C.-H.; Wu, P.-C.; Lee, W. Polymer Stabilization of Uniform Lying Helix Texture in a Bimesogen-Doped Cholesteric Liquid Crystal for Frequency-Modulated Electro-Optic Responses. *Materials* **2022**, *15*, 771. [CrossRef] [PubMed]
7. Bikbaev, R.G.; Maksimov, D.N.; Chen, K.-P.; Timofeev, I.V. Double-Resolved Beam Steering by Metagrating-Based Tamm Plasmon Polariton. *Materials* **2022**, *15*, 6014. [CrossRef] [PubMed]
8. Reshetnyak, V.Y.; Pinkevych, I.P.; Bunning, T.J.; Evans, D.R. Influence of Rugate Filters on the Spectral Manifestation of Tamm Plasmon Polaritons. *Materials* **2021**, *14*, 1282. [CrossRef] [PubMed]
9. Avdeeva, A.Y.; Vetrov, S.Y.; Bikbaev, R.G.; Pyatnov, M.V.; Rudakova, N.V.; Timofeev, I.V. Chiral Optical Tamm States at the Interface between a Dye-Doped Cholesteric Liquid Crystal and an Anisotropic Mirror. *Materials* **2020**, *13*, 3255. [CrossRef] [PubMed]
10. Lin, M.-Y.; Xu, W.-H.; Bikbaev, R.G.; Yang, J.-H.; Li, C.-R.; Timofeev, I.V.; Lee, W.; Chen, K.-P. Chiral-Selective Tamm Plasmon Polaritons. *Materials* **2021**, *14*, 2788. [CrossRef] [PubMed]

Article

Smart Window with Active-Passive Hybrid Control

Heng-Yi Tseng [1], Li-Min Chang [1], Kuan-Wu Lin [1], Cheng-Chang Li [1], Wan-Hsuan Lin [1], Chun-Ta Wang [1], Chien-Wen Lin [2], Shih-Hsien Liu [2] and Tsung-Hsien Lin [1,*]

[1] Department of Photonics, National Sun Yat-sen University, Kaohsiung 80424, Taiwan; coololiver0130@gmail.com (H.-Y.T.); book74108520@gmail.com (L.-M.C.); kobe0811219@gmail.com (K.-W.L.); ronnie.ccl8002@gmail.com (C.-C.L.); tiffanywanhsuan@gmail.com (W.-H.L.); wangchunta4@gmail.com (C.-T.W.)

[2] Material and Chemical Research Laboratories of Industrial Technology Research Institute, Hsinchu 31040, Taiwan; linjianwen@itri.org.tw (C.-W.L.); shliu@itri.org.tw (S.-H.L.)

* Correspondence: jameslin@mail.nsysu.edu.tw; Tel.: +886-07-525-2000 (ext. 4442)

Received: 26 August 2020; Accepted: 14 September 2020; Published: 17 September 2020

Abstract: Dimming and scattering control are two of the major features of smart windows, which provide adjustable sunlight intensity and protect the privacy of people in a building. A hybrid photo- and electrical-controllable smart window that exploits salt and photochromic dichroic dye-doped cholesteric liquid crystal was developed. The photochromic dichroic dye causes a change in transmittance from high to low upon exposure to sunlight. When the light source is removed, the smart window returns from colored to colorless. The salt-doped cholesteric liquid crystal can be bi-stably switched from transparent into the scattering state by a low-frequency voltage pulse and switched back to its transparent state by a high-frequency voltage pulse. In its operating mode, an LC smart window can be passively dimmed by sunlight and the haze can be actively controlled by applying an electrical field to it; it therefore exhibits four optical states—transparent, scattering, dark clear, and dark opaque. Each state is stable in the absence of an applied voltage. This smart window can automatically dim when the sunlight gets stronger, and according to user needs, actively adjust the haze to achieve privacy protection.

Keywords: smart window; cholesteric liquid crystal; photochromic dichroic dye

---

1. Introduction

Global warming has become the most important environmental issue. Smart windows are very important in energy-saving buildings, where they save energy and contribute to the reduction of carbon emissions [1–3]. A smart window allows the transmission of light to be adjusted; it can protect against sunlight and thermally insulate, effectively reducing the need for lighting and air conditioning. Liquid crystal has been used to manufacture various light shutters and switchable windows. Methods for controlling liquid crystal smart windows are divided into two types—electrical control and photo control. The advantage of an electrical-controllable liquid crystal smart window is that the user can freely set the transmittance or the degree of transmittance by adjusting the applied voltage (see, for example, the dynamic scattering mode [4,5], polymer-dispersed liquid crystals [6–9], polymer-stabilized cholesteric texture [10–12], and dye-doped liquid crystals [9,13]). Since the general nematic liquid crystal has dielectric anisotropy, its liquid molecules rotate in an applied electric field, so electrical control is an ordinary driving method for a liquid crystal smart window. The advantage of a photo-controllable liquid crystal smart window is that it can automatically adjust transmittance from high to low upon exposure to UV, as in sunlight, and so self-adjust without extra energy input. The photo-controllable liquid crystal smart window requires additional photo-sensitive material (such as photosensitive chiral azobenzene [14,15], photochromic dye [16], or azo dye [17]). Most of the photo-controlled liquid crystal smart windows perform the sole function of either absorption or scattering. A single device with multiple states to meet different haze and tint needs is highly desired but challenging to develop. Most liquid crystal smart windows cannot have both photo- and electrical-controllable capabilities, and their transmittance cannot be adjusted to meet user needs.

Cholesteric liquid crystal (CLC) is a smart window material that exhibits bi-stability [18,19]. This bi-stable characteristic eliminates the need for the continuous application of a voltage, as is required with ordinary smart windows. A voltage pulse has only to be applied when switching, so power consumption is very low. CLC has two stable states, which are associated with different optical properties. The first state is the planar state [18], in which CLC virtually functions as a chiral Bragg grating with a photonic bandgap that is centered at wavelength $\lambda_c = (n_o + n_e)p/2$ and has band width $\Delta\lambda \approx (n_o - n_e)p$, where n_o is the ordinary refractive index, n_e is the extraordinary index, and p is the helical pitch. The band gap is easily tunable by adjusting the concentration of the chiral agent. Inside the band gap, the incident light is reflected by the CLC; outside the band gap, the light directly passes through the CLC without scattering. The other stable state is the focal conic state, in which the helical axes are randomly distributed, forming a multi-domain structure with strong light scattering. The CLC must generally have a particular cell gap-to-pitch ratio (d/p) and surface alignment to maintain its stability [13,20]. In a vertical alignment film with large d/p, the cholesteric liquid crystal cannot be easily be put into the planar state, disfavoring transparency; in a homogenous alignment film with small d/p, the cholesteric liquid crystal cannot easily enter the focal conic state, disfavoring scattering and resulting in a poor contrast ratio. In recent researches, a cholesteric liquid crystal light shutter has been developed. This shutter has a cholesteric liquid crystal with negative dielectric anisotropy; it is switched by a high-frequency voltage pulse switched into the transparent planar state and by a low-frequency voltage pulse back into the scattering focal conic state [21–23]. This operation increases the tolerance of the CLC for d/p and boundary conditions, and a high-frequency voltage can be used to obtain good planar states even when the cell gap is large or pitch is short (large d/p).

In this work, a photochromic dye is added to a salt-doped cholesteric liquid crystal. The photochromic dye is photo-controllable and automatically darkens when irradiated by UV light. It can regulate the transmittance without sacrificing transparency. The electrical controllability of the CLC is used to control the translucency of smart windows to provide privacy protection. Finally, four optical states are realized—transparent, scattering, dark clear (chromatic and transparent), and dark opaque (chromatic and hazy).

2. Materials and Methods

2.1. Preparation of Materials and Measurements

The cholesteric liquid crystal (CLC) mixture was made from 96 wt.% negative dielectric anisotropy nematic liquid crystal DNM-9528 (extraordinary index $n_e \approx 1.5792$, ordinary index $n_o \approx 1.4808$, dielectric anisotropy $\Delta\varepsilon = \varepsilon_{//} - \varepsilon_{\perp} \approx -4.8$, rotational viscosity coefficient $\gamma1 = 96$ mPa·s, splay elastic constant K11 = 12.4 pN and bend elastic constant K33 = 12.8 pN), doped with 4 wt.% chiral agent R-5011 (HTP ≈ 100 μm^{-1}, from HCCH, Jiangsu, China), whose chemical structure is presented in Figure 1; the pitch of the CLC was approximately 250 nm. Then, 97.65 wt.% CLC mixture was further homogeneously blended with 0.35 wt.% tetrabutylammonium tetrafluoroborate $((CH_3CH_2CH_2CH_2)_4N(BF_4))$ salt (TBATFB, from Sigma Aldrich, St. Louis, MO, United States), and 2 wt.% photochromic dichroic dye molecule ethyl 8-((4'-pentylcyclohexylphenyl)-difluoromethylphenyl-4-yl)-2-phenyl-2-(4-pyrrolidinyl phenyl)-2H-naphtho[1,2-b]pyran-5-carboxylate (from ITRI, Hsinchu, Taiwan), the chemical structure presented in Figure 2. All of the mixture was sandwiched between two glass substrates without any alignment treatment; the inner surfaces of both substrates were pre-coated with indium tin oxide to cause them to function as transparent conductive layers. The cell gap was approximately 12 μm.

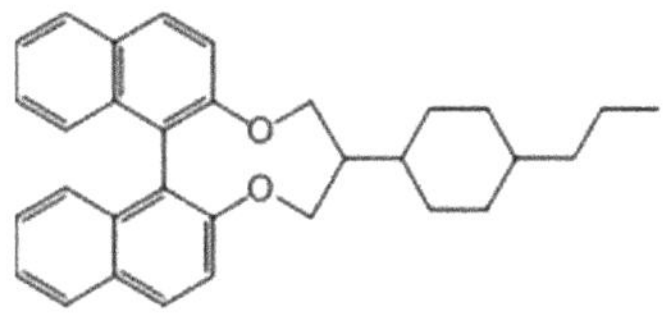

Figure 1. Chemical structure of chiral agent R-5011.

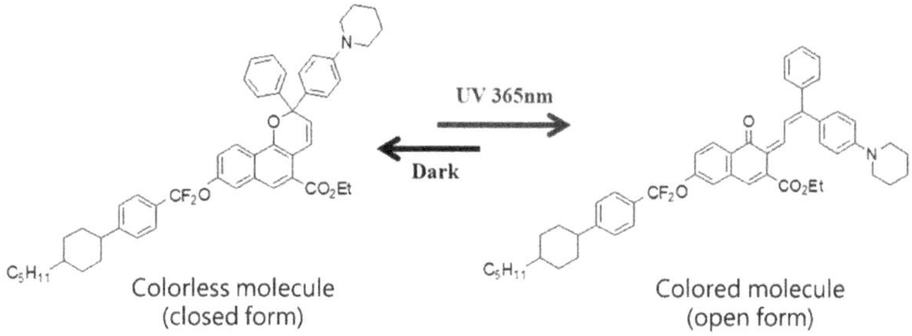

Figure 2. Chemical structure of photochromic dichroic dye ethyl 8-((4'-pentylcyclohexylphenyl)-difluoromethylphenyl-4-yl)-2-phenyl-2-(4-pyrrolidinyl phenyl)-2H-naphtho[1,2-b]pyran-5-carboxylate.

Spectra were obtained using a USB4000 spectrometer (from Ocean Optics, Largo, FL, USA). The source of the incident light was a parallel tungsten halogen light HL-2000 (from Ocean Optics). The haze was measured using a haze meter NDH2000 (from Nippon Denshoku, Tokyo, Japan).

2.2. Driving Method and Principle

The smart window herein can be made to enter four optical states by exposure to UV light and the application of various voltages. Figure 3 shows a schematic switching diagram. In the absence of UV light, the sample is initially in the planar state, in which the liquid crystal molecules are all parallel to the substrate; the reflection wavelength of this cholesteric liquid crystal is modulated in the UV light region, so it is transparent in the visible light region, as shown in Figure 3a. Since the salt TBATFB is doped into the CLC, it disassociates into positive and negative ions and changes the electric conductivity of the liquid crystal. When a low-frequency voltage is applied to an LC with negative dielectric anisotropy ($\Delta\varepsilon < 0$) and positive conductivity anisotropy ($\Delta\sigma > 0$), the turbulence effect can be observed. The ions will move parallel to the electric field [24]. The motion of the ions creates turbulence and tends to align the liquid crystal along the direction of movement. This turbulence effect distorts the helical structure in the planar state relative to that in the focal conic state. When the electric field is removed, the cholesteric liquid crystal remains in the focal conic state. The focal conic state displayed scattering because of its multi-domain structure, as shown in Figure 3b. To switch the CLC from the focal conic state to the planar state, a high-frequency voltage is applied. Because of the limited mobility of the ions, the high-frequency electric field cannot generate turbulence. Owing to the negative dielectric anisotropy of the cholesteric liquid crystal, the electric field tends to align the liquid crystal molecules perpendicular to the electric field. The helical twisting power that is generated by the chirality causes the CLC to enter the planar state. Therefore, the CLC can be switched between the planar and focal conic states by applying high-frequency and low-frequency pulse electric fields. Both of these states are stable at 0 V, in which the material is transparent and scattering, respectively.

The smart window is colorless when it is not irradiated by UV, but rapidly darkens when it is irradiated by UV. Photochromic dichroic dye naphthopyran-based materials undergo photo-induced conformational changes in molecular structure from closed form to open form. The structure of the dichroic dye is elongated and planar in its open form. The absorbance spectra of dichroic dye undergoes a red-shift from ultraviolet to the visible range upon UV light irradiation. Photo-induced conformational changes are short-lived and reversible. When the UV light source is removed, the material spontaneously returns to its closed, colorless form in a few minutes. Since the helical structure in the planar state has the helical axes perpendicular to the substrate, incident light of any polarization will be strongly absorbed by the photochromic dichroic dye, reducing transmittance; this state is the dark clear state, as shown in Figure 3c. Similarly, in the dark clear state, a low-frequency pulse voltage can be applied to cause ionic turbulence and formed focal conic state. In this state, the absorption by the doped photochromic dichroic dye and multi-domain scattering characteristics cause the smart window to be dark and opaque; in this state, the smart window has the lowest transmittance, as shown in Figure 3d. The smart window becomes colorless after the UV irradiation has been removed for a short period.

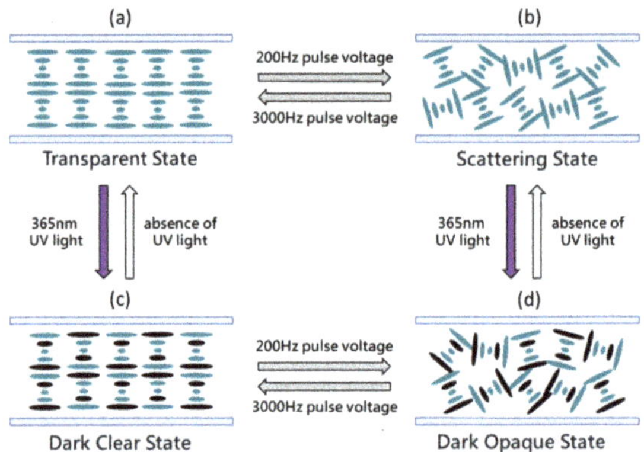

Figure 3. Schematic diagram and operation of photo- and electrical-controllable smart window. (**a**) Transparent planar state in visible region. (**b**) Scattering focal conic state. (**c**) Dimming (absorption) planar state. (**d**) Absorption scattering focal conic state.

3. Results

The degree of absorption of visible light by photochromic dichroic dye can be controlled by exposure to, or the absence of, UV light. Scattering can be controlled by applying voltage pulses of different frequencies. Therefore, this smart window could adopt four optical states—transparent, scattering, dark clear, and dark opaque; each state was stable at 0 V. Figure 4 shows the transmission spectra in these four states, which are compared in Table 1 with respect to average transmittance and haze. The transmittance in the transparent state was high in visible region and almost no scattering occurred; the average transmittance exceeded 83%, and the haze was less than 6%; in the scattering state, the haze was uniform; the average transmittance was only 32.2% and the haze as high as 70.4%. When the dye was irradiated for 1 min with UV light with a wavelength of 365 nm and an intensity of 10 mw/cm^2, the dye photochromic dichroic transitioned from closed form to open form and absorbed. Since the absorption peak of the photochromic dye is around 560 nm, the transparent state had deep purple discoloration. The average transmittance of the dark clear state was 30.3%, and the haze was 8.3%, slightly higher than in the transparent state; in the dark clear state, the average transmittance was lowest at 14.2%, and the haze was highest at 74.9%.

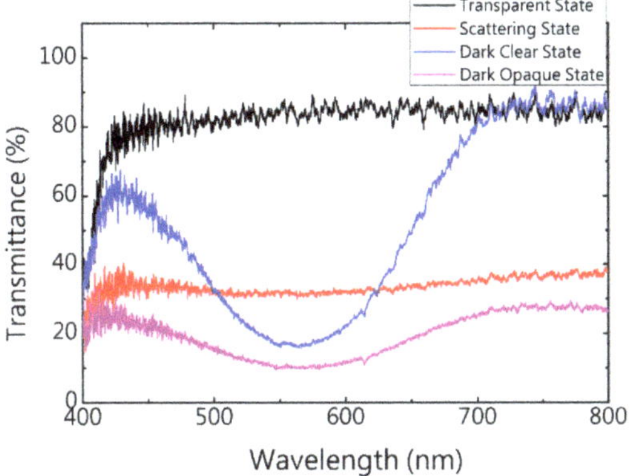

Figure 4. Transmission spectra in different optical states.

Table 1. Comparisons of optical states.

Optical Properties	Transparent State	Scattering State	Dark Clear State	Dark Opaque State
Average Transmittance (%)	83.35	32.3	30.3	14.2
Haze (%)	5.9	70.4	8.3	74.9

Transmittance versus frequency of voltage pulse is shown in the Figure 5. The CLC was initially in the planar state with high transmittance, then a 60 V voltage pulse with different frequency was applied to measure the transmittance. When a low-frequency voltage was applied across the cell, the motion of the ions created turbulence and tended to align the liquid crystal along the direction of movement. This turbulence effect distorted the helical structure and turned the CLC from the focal conic state to the planar state, where we chose 200 Hz as the low-frequency voltage pulse. When the frequency increased, the turbulence became weak and the dielectric effect of LC dominated the alignment. Therefore, it maintained in the planar state with high transmittance. Here we chose 3000 Hz as the high-frequency voltage pulse.

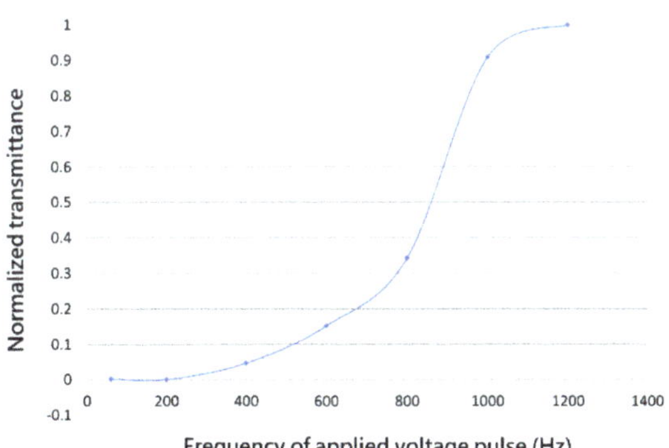

Figure 5. Normalized transmission as a function of the frequency of applied voltage (60 V).

The electro-optical characteristic of the smart window was measured before and after irradiation by UV light; a voltage pulse with a duration time of 3 s was applied and the transmission was measured 10 s after the end of the pulse. Hence, all transmissions were measured at 0 V. Figure 6 plots the transmittance of the smart window after the removal of the applied voltage vs. the amplitude of the applied voltage pulse. A parallel tungsten halogen was the light source and a spectrometer was used as the detector. Figure 6a plots the T–V curve of the smart window in the absence of UV light. The transparent state switched into the scattering state as shown in Figure 6a (black line). Voltage pulses with a frequency of 200 Hz switched the liquid crystal from the transparent planar state to the scattering focal conic state. When the voltage pulse was below 40 V, the turbulence that was generated by the ions was not enough to cause a transition of the liquid crystal; when a voltage pulse over 40 V was applied, the turbulence effect was against the dielectric interaction because of the negative dielectric anisotropy of the liquid crystal; the helical structure in the planar state was disturbed, and ultimately switched to the focal conic state. As the voltage was increased, the turbulence effect became stronger, and the focal conic domain size became smaller, close to the wavelength of visible light. Therefore, more scattering reduced transmittance. When a voltage pulse of 50 V was applied, the transmittance was at its minimum, and the ionic turbulence was greater than the dielectric interaction. The scattering then switched into the transparent state as shown in Figure 6a (red line). Voltage pulses with a frequency of 3000 Hz switched the liquid crystal from the focal conic into the planar state. Pulses of 50 V and a frequency of 200 Hz were applied to reset the sample into the focal conic state before applied measurement voltage pulses. A voltage pulse with a duration time of 3 s was applied and the transmission was measured 10 s after the end of the pulse. Voltage pulses with a frequency of 3000 Hz yielded very weak ionic turbulence because the high-frequency voltage limited their mobility, so only negative dielectric anisotropy contributed to all of the electrical properties. When the voltage pulse was less than 20 V, the liquid crystal molecules were not arranged in parallel as the substrate, and the focal conic state was maintained. As the applied voltage pulse increased, some regions adopted the parallel substrate alignment due to the negative dielectric anisotropy of liquid crystal molecules and the formation of a planar structure as a result of the helical twisting power that arose from the chiral doping. When the voltage pulse exceeded 70 V, all liquid crystals were arranged in a planar state as a result of the electric field, and this state was stably maintained after the voltage was turned off.

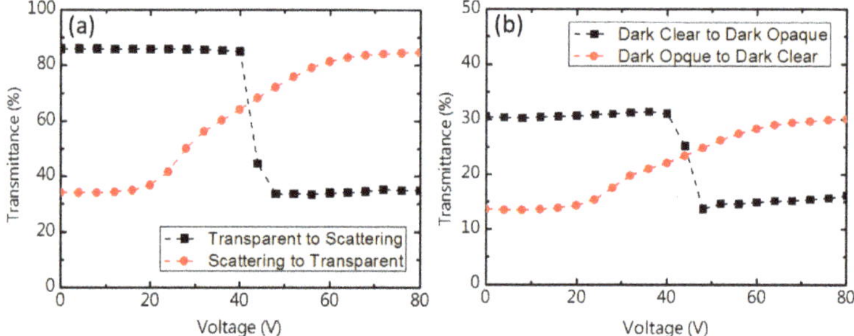

Figure 6. Transmittance versus amplitude of voltage pulse (**a**) before irradiation by UV light and (**b**) after irradiation by UV light.

Figure 6b plots the T–V curve of the smart window that was exposed to UV light. The sample was exposed for 1 min to UV light with a wavelength of 365 nm and an intensity 10 mw/cm^2 before the transmittance was measured as a function of the amplitude of the voltage pulse. Irradiation by UV light changed the sample from colorless to deep purple. Owing to the structure in the planar state and the dichroic absorption of the photochromic dye, the transmittance decreased to about 30%. The focal conic state was dark and hazy with a lower transmittance of 14.2%. Switching completely from the dark clear state to the dark opaque state required a 52 V voltage pulse with the frequency of 200 Hz. Switching from the dark opaque state to the dark clear state required a 70 V voltage pulse with a frequency of 3000 Hz.

Table 2 compares the response time before and after irradiated by UV light. A 52 V voltage pulse with a frequency of 200 Hz was used to switch from the planar state to the focal conic state before and after irradiation by UV light. The response times before and after the color changed were 292 ms and 277 ms, respectively. A 70 V voltage pulse with a frequency of 3000 Hz changed the focal conic state to the planar state. The response times before and after the color changed were 16.8 ms and 23.6 ms, respectively. When the smart window was switched from the transparent state to the scattering state, the competition between the turbulence effect and the negative dielectric anisotropy of the liquid crystal caused the response time to be much longer than that associated with switching from the scattering state to the transparent state. Based on the above results of the experiments on the electro-optical characteristics, both in the presence and the absence of UV light, the photochromic dichroic dye only affects the optical properties, whereas the electrical properties do not significantly change.

Table 2. Comparisons of response time before and after irradiation by UV light.

Response Time	Form Transparent to Scattering	Form Scattering to Transparent
Before Color Change (ms)	292	16.8
After Color Change (ms)	277.2	23.6

Figure 7 plots the time-dependence of the optical response to exposure UV light. The smart window in the planar state was irradiated by UV light with a wavelength of 365 nm and an intensity of 10 mw/cm^2 for 42 s, and thus changed from colorless to deep purple. The response time for the normalized transmittance to decline from 90% to 10% was 5.5 s. The UV light was then turned off and the light transmittance of the smart window returned to high. The response time for the normalized transmittance to increase from 10% to 90% was 53.9 s.

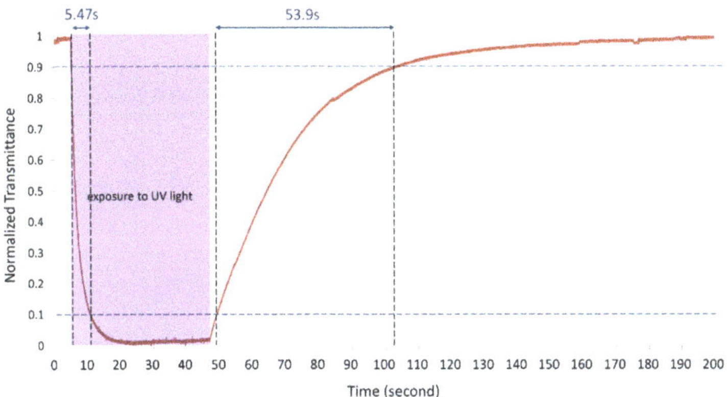

Figure 7. Optical response time under irradiation by UV light with wavelength of 365 nm and an intensity of 10 mw/cm^2.

Figure 8 presents photographs of the proposed smart window. Each state existed stably without the application of a field. Figure 8a,b show the transparent state and scattering state without exposure to UV light. The transparent state is colorless and allows the background to be seen clearly. The scattering state is white and blocks the background. Figure 8c,d present the dark clear state and the dark opaque state under irradiation by UV light. In the dark clear state, since the absorption peak of the photochromic dye is around 560 nm, the sample has deep purple discoloration. The dark opaque state exhibits absorption and scattering, and so has the lowest transmittance and provides the best shielding. Figure 9 shows microscopic images of each optical states.

Figure 10 shows photographs of this smart window irradiated by real sunlight at noon at National Sun Yat-sen University. Before it is exposed to sunlight, the sample is in the transparent planar state and colorless, as displayed in Figure 10a. Figure 10b shows this smart window covered by a checkerboard mask and exposed to sunlight for five minutes. The part that was irradiated by sunlight entered the dark clear state. Finally, the mask was removed and the sample was exposed to sunlight for five minutes; a uniform dark clear state was obtained across the entire surface, as shown in Figure 10c.

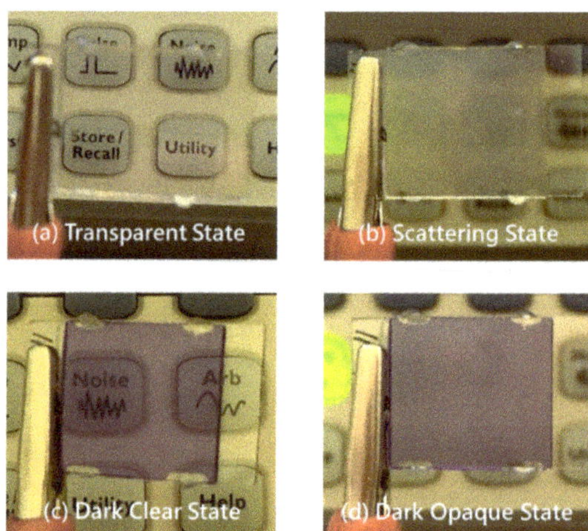

Figure 8. Photographs of (**a**) transparent state, (**b**) scattering state, (**c**) dark clear state, and (**d**) dark opaque state.

Figure 9. Microscopic images of (**a**) transparent state, (**b**) scattering state, (**c**) dark clear state, and (**d**) dark opaque state.

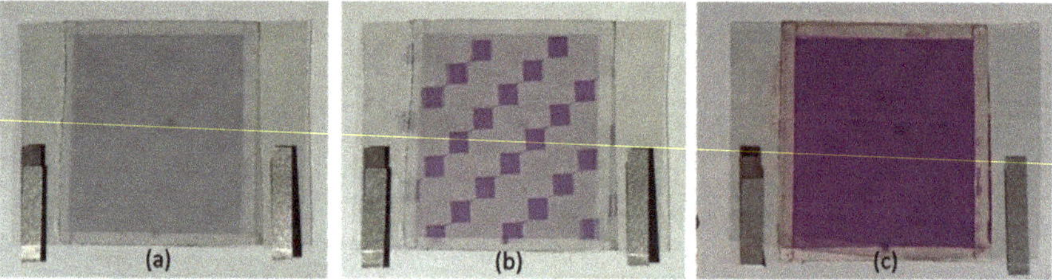

Figure 10. Photographs of this smart window (**a**) in planar state before exposure to sunlight, (**b**) covered by a checkerboard mask and exposed to sunlight for five minutes, (**c**) after the removal of the mask and exposed to sunlight for five minutes.

4. Conclusions

In summary, this work demonstrated a photochromic dye and salt-doped cholesteric liquid crystal smart window that has four different optical states—transparent, scattering, dark clear, and dark opaque. The smart window can be switched from the transparent state (planar state) to the scattering state (focal conic state) by applying a 52 V voltage pulse with a frequency of 200 Hz, causing the transmittance to decrease from 83.35% to 32.3%; it can be switched back to the transparent planar state by applying a 70 V voltage pulse with a frequency of 3000 Hz, each state is stable over 48 hours or even longer in the absence of an applied voltage. Exposure to natural sunlight or artificial UV light changes its chromatic properties. The dimming transmittance is 30.3% and 14.2% in the dark and dark opaque states, respectively. After the UV source is removed, the smart window returns to its colorless state. Therefore, this smart window, which provides passive automatic dimming and active haze control for privacy protection, has great practical potential.

Author Contributions: Conceptualization, C.-T.W., S.-H.L. and T.-H.L.; Formal analysis, H.-Y.T., L.-M.C., K.-W.L. and C.-W.L.; Data curation, H.-Y.T., L.-M.C., K.-W.L. and W.-H.L.; Investigation, H.-Y.T., C.-C.L. and T.-H.L.; Methodology, H.-Y.T., C.-C.L. and C.-W.L.; Supervision, C.-T.W., S.-H.L. and T.-H.L.; Project administration, C.-T.W., S.-H.L. and T.-H.L.; Writing—original draft, H.-Y.T. All authors have read and agreed to the published version of the manuscript.

Funding: This research was funded by the Ministry of Science and Technology, Taiwan, under contract no. MOST 106-2112-M-110-003-MY3, MOST 107-2628-E-110-001-MY2 and MOEA (Taiwan) project E301ARY4N1 implemented by MCL/ITRI.

Acknowledgments: The authors would like to thank the Ted Knoy for his editorial assistance.

Conflicts of Interest: The authors declare no conflict of interest.

References

1. Baetens, R.; Jelle, B.P.; Gustavsen, A. Properties, requirements and possibilities of smart windows for dynamic daylight and solar energy control in buildings: A state-of-the-art review. *Sol. Energy Mater. Sol. Cells* **2010**, *94*, 87–105. [CrossRef]

2. Aburas, M.; Soebarto, V.; Williamson, T.; Liang, R.; Ebendorff-Heidepriem, H.; Wu, Y. Thermochromic smart window technologies for building application: A review. *Appl. Energy* **2019**, *255*, 113522. [CrossRef]

3. Lampert, C.M. Smart switchable glazing for solar energy and daylight control. *Sol. Energy Mater. Sol. Cells* **1998**, *52*, 207–221. [CrossRef]

4. Heilmeier, G.H.; Zoanoni, L.A.; Barton, L.A. Dynamic Scattering in Nematic Liquid Crystals. *Appl. Phys. Lett.* **1968**, *13*, 46. [CrossRef]

5. Zhan, Y.; Lu, H.; Jin, M.; Zhou, G. Electrohydrodynamic instabilities for smart window applications. *Liq. Cryst.* **2020**, *47*, 977–983. [CrossRef]

6. Bouteiller, L.; Barny, P.L. Polymer-dispersed liquid crystals: Preparation, operation and application. *Liq. Cryst.* **1996**, *21*, 157–174. [CrossRef]

7. Kim, M.; Park, K.J.; Seok, S.; Ok, J.M.; Jung, H.-T.; Choe, J.; Kim, D.H. Fabrication of Microcapsules for Dye-Doped Polymer-Dispersed Liquid Crystal-Based Smart Windows. *ACS Appl. Mater. Interfaces* **2015**, *7*, 17904–17909. [CrossRef]

8. Pane, S.; Caporusso, M.; Hakemi, H. Haze and opacity control in polymer dispersed liquid crystal (PDLC) films with phase separation method. *Liq. Cryst.* **1997**, *23*, 861–867. [CrossRef]

9. Oh, S.-W.; Baek, J.-M.; Heo, J.; Yoon, T.-H. Dye-doped cholesteric liquid crystal light shutter with a polymer-dispersed liquid crystal film. *Dye. Pigment.* **2016**, *134*, 36–40. [CrossRef]

10. Bao, R.; Liu, C.-M.; Yang, D.-K. Smart bistable polymer stabilized cholesteric texture light shutter. *Appl. Phys. Express* **2009**, *2*, 112401. [CrossRef]

11. Ma, J.; Shi, L.; Yang, D.-K. Bistable polymer stabilized cholesteric texture light shutter. *Appl. Phys. Express* **2010**, *3*, 21702. [CrossRef]

12. Li, C.-C.; Tseng, H.-Y.; Pai, T.-W.; Wu, Y.-C.; Hsu, W.-H.; Jau, H.-C.; Chen, C.-W.; Lin, T.-H. Bistable cholesteric liquid crystal light shutter with multielectrode driving. *Appl. Opt.* **2014**, *53*, E33–E37. [CrossRef] [PubMed]

13. Yu, B.-H.; Huh, J.-W.; Kim, K.-H.; Yoon, T.-H. Light shutter using dichroic-dye-doped long-pitch cholesteric liquid crystals. *Opt. Express* **2013**, *21*, 29332–29337. [CrossRef]

14. Wang, C.-T.; Wu, Y.-C.; Lin, T.-H. Photo-controllable tristable optical switch based on dye-doped liquid crystal. *Dye. Pigment.* **2014**, *103*, 21–24. [CrossRef]

15. Talukder, J.R.; Lee, Y.H.; Wu, S.T. Photo-responsive dye-doped liquid crystals for smart windows. *Opt. Express* **2019**, *27*, 4480–4487. [CrossRef]

16. Macchione, M.; De Filpo, G.; Nicoletta, F.P.; Chidichimo, G. Photochromic reverse mode polymer dispersed liquid crystals. *Liq. Cryst.* **2005**, *32*, 315–319. [CrossRef]

17. Goda, K.; Omori, M.; Takatoh, K. Optical switching in guest–host liquid crystal devices driven by photo- and thermal isomerisation of azobenzene. *Liq. Cryst.* **2018**, *45*, 485–490. [CrossRef]

18. Yang, D.K.; West, J.L.; Chien, L.C.; Doane, J.W. Control of reflectivity and bistability in displays using cholesteric liquid crystals. *J. Appl. Phys.* **1994**, *76*, 1331–1333. [CrossRef]

19. Chen, C.-W.; Brigeman, A.N.; Ho, T.-J.; Khoo, I.C. Normally transparent smart window based on electrically induced instability in dielectrically negative cholesteric liquid crystal. *Opt. Mater. Express* **2018**, *8*, 691–697. [CrossRef]

20. Li, C.-C.; Tseng, H.-Y.; Liao, H.-C.; Chen, H.-M.; Hsieh, T.; Lin, S.-A.; Jau, H.-C.; Wu, Y.-C.; Hsu, Y.-L.; Hsu, W.-H. Enhanced image quality of OLED transparent display by cholesteric liquid crystal back-panel. *Opt. Express* **2017**, *25*, 29199–29206. [CrossRef]

21. Moheghi, A.; Nemati, H.; Li, Y.; Li, Q.; Yang, D.-K. Bistable salt doped cholesteric liquid crystals light shutter. *Opt. Mater.* **2016**, *52*, 219–223. [CrossRef]

22. Lan, Z.; Li, Y.; Dai, H.; Luo, D. Bistable Smart Window Based on Ionic Liquid Doped Cholesteric Liquid Crystal. *IEEE Photonics J.* **2017**, *9*, 1–7. [CrossRef]

23. Cheng, K.-T.; Lee, P.-Y.; Qasim, M.M.; Liu, C.-K.; Cheng, W.-F.; Wilkinson, T.D. Electrically Switchable and Permanently Stable Light Scattering Modes by Dynamic Fingerprint Chiral Textures. *ACS Appl. Mater. Interfaces* **2016**, *8*, 10483–10493. [CrossRef] [PubMed]

24. Chiang, S.-P.; Wang, C.-T.; Feng, T.-M.; Li, C.-C.; Jau, H.-C.; Su, S.-Y.; Yang, S.-D.; Lin, T.-H. Selective variable optical attenuator for visible and mid-Infrared wavelengths. *Opt. Express* **2018**, *26*, 17009–17014. [CrossRef]

Article

Effect of Host Structure on Optical Freedericksz Transition in Dye-Doped Liquid Crystals

Junki Yokota [1,2], Kohsuke Matsumoto [1,2], Koji Usui [1,2], Shoichi Kubo [1,2] and Atsushi Shishido [1,2,*]

1   Laboratory for Chemistry and Life Science, Institute of Innovative Research, Tokyo Institute of Technology, R1-12, 4259 Nagatsuta, Midori-ku, Yokohama 226-8503, Japan; jyokota-st@polymer.res.titech.ac.jp (J.Y.); kmatsumoto-st@polymer.res.titech.ac.jp (K.M.); kusui-st@polymer.res.titech.ac.jp (K.U.); kubo@res.titech.ac.jp (S.K.)

2   Department of Chemical Science and Engineering, School of Materials and Chemical Technology, Tokyo Institute of Technology, 2-12-1 Ookayama, Meguro-ku, Tokyo 152-8552, Japan

*   Correspondence: ashishid@res.titech.ac.jp; Tel.: +81-45-924-5242

Abstract: The optical Freedericksz transition (OFT) can reversibly control the molecular orientation of liquid crystals (LCs) only by light irradiation, leading to the development of all-optical devices, such as smart windows. In particular, oligothiophene-doped LCs show the highly sensitive OFT due to the interaction between dyes and an optical-electric field. However, the sensitivity is still low for the application to optical devices. It is necessary to understand the factors in LCs affecting the OFT behavior to reduce the sensitivity. In this study, we investigated the effect of the host LC structure on the OFT in oligothiophene-doped LCs. The threshold light intensity for the OFT in trifluorinated LCs was 42% lower than that in LCs without fluorine substituents. This result contributes to the material design for the low-threshold optical devices utilizing the OFT of dye-doped LCs.

Keywords: optical Freedericksz transition; dye-doped liquid crystal; molecular reorientation

Citation: Yokota, J.; Matsumoto, K.; Usui, K.; Kubo, S.; Shishido, A. Effect of Host Structure on Optical Freedericksz Transition in Dye-Doped Liquid Crystals. *Materials* **2022**, *15*, 4125. https://doi.org/10.3390/ma15124125

Academic Editors: Wei Lee and Ivan V. Timofeev

Received: 8 April 2022
Accepted: 8 June 2022
Published: 10 June 2022

Publisher's Note: MDPI stays neutral with regard to jurisdictional claims in published maps and institutional affiliations.

1. Introduction

The Freedericksz transition is a phenomenon to reversibly change the molecular orientation of liquid crystals (LCs) by external stimuli, such as an electric, magnetic, and optical-electric field above a certain intensity. This phenomenon is applicable to functional optical devices, such as displays and smart windows [1–3]. In particular, the optical Freedericksz transition (OFT) induced only by an optical-electric field has attracted much attention due to the potential for the development of energy-saving devices driven by sunlight. However, the OFT still requires high light intensity above 10^2 W/cm^2 [4–8]. Therefore, enhancing the sensitivity of OFT, that is, the sensitivity of the molecular reorientation of LC is essential for the application to all-optical devices.

One approach to improve the sensitivity of the molecular reorientation is doping the LCs with small amounts of certain dichroic organic dyes. This approach was first reported by Jánossy et al., where the sensitivity of the molecular reorientation of LCs was enhanced by the addition of anthraquinone dyes, resulting in inducing the OFT at a ten-times lower light intensity than LCs without dyes [9–11]. The rod-like dyes excited by the irradiation with a linearly polarized laser beam are aligned parallel to the incident polarization direction due to the interaction with the optical-electric field. Consequently, LC molecules are also reoriented along the same direction as the dye molecules due to the cooperative interaction.

Instead of the anthraquinone dyes, Zhang et al. reported the usage of oligothiophene dyes to enhance the sensitivity of the LC molecular reorientation [12]. To further decrease the light intensity required for the OFT in oligothiophene-doped LCs, various approaches focusing on the incident light, alignment processing, and materials have been proposed [13–19]. Recently, we focused on the material design and achieved the effective molecular reorientation by doping oligothiophene dyes with ester moieties [17,18]. In addition, we found that incorporating a polymer into dye-doped LCs caused the enhancement and stabilization of the molecular reorientation compared to conventional low-molecular-weight LCs [19,20]. However, the effect of host LCs, which determines the physical properties of the LC systems, on the OFT remained unexplored in oligothiophene dye-doped LCs. Investigation of the effect of host LCs gives an insight into the material design for enhancing the sensitivity of the OFT.

In this work, we explored the OFT behavior in oligothiophene dye-doped LCs using fluorinated host LCs in terms of the effect of the electron withdrawing group. The OFT was measured by observing the diffraction rings caused by the molecular reorientation of LCs. The light intensity at which the molecular reorientation occurred depended on the structure of the host LCs. The dye-doped LCs using host LCs with three fluorine substituents induced the OFT at a lower light intensity than previously reported dye-doped LC. Furthermore, the dye-doped LCs containing host LCs with two fluorine substituents required a higher light intensity than those with three fluorine substituents. We revealed that the number of fluorine substituents of host LCs greatly affected the sensitivity of the OFT. This result can provide important knowledge for designing highly sensitive dye-doped LCs.

2. Materials and Methods

2.1. Sample Preparation

Figure 1 shows the chemical structures of the compounds used in this study. As host LCs, we used 4-cyano-4'-pentyl biphenyl (5CB) obtained from Merck Ltd., Tokyo, Japan, and four types of fluorinated LCs (F-LCx, x = 1–4) obtained from Tokyo Chemical Industry Co. Ltd., Tokyo, Japan. F-LCx are as follows: *trans*-4-(3,4-difluorophenyl)-*trans*-4'-ethylbicyclohexane (F-LC1); *trans,trans*-4'-ethyl-4-(3,4,5-trifluorophenyl)bicyclohexyl (F-LC2); 3,4-difluoro-4'-(*trans*-4-propylcyclohexyl)biphenyl (F-LC3); and 3,4,5-trifluoro-4'-(*trans*-4-propylcyclohexyl)biphenyl (F-LC4). A dichroic oligothiophene dye molecule, 5,5''-bis-(5-butyl-2-thienylethynyl)-2,2':5',2''-terthiophene (TR5), was synthesized as reported previously [12].

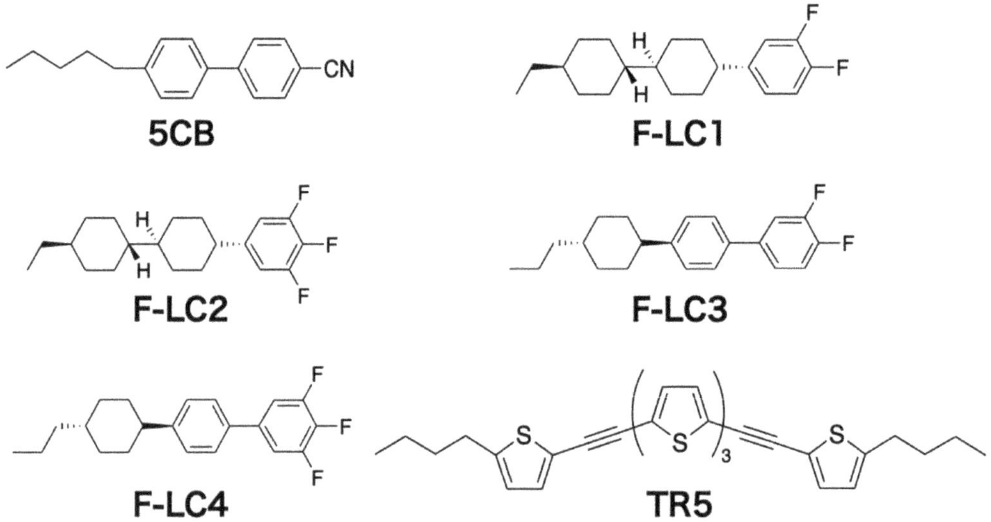

Figure 1. Chemical structures of the compounds used in this study.

Pure 5CB or mixtures of 5CB and F-LCx (5CB:F-LCx = 75:25, molar ratio) were mixed with a TR5 solution (0.01 mol/L in tetrahydrofuran (THF)) in brown vials, diluted in THF, and stirred for 1 h. After stirring, the solvent was completely removed under vacuum. The TR5 concentration was adjusted for each host mixture to show the same absorbance. The details are described in the "Result & Discussion" section.

Glass cells with homeotropic alignment layers were fabricated according to the procedure shown in Figure 2. Glass substrates (25 mm × 25 mm) ultrasonicated in 2-propanol were treated with an ultraviolet (UV)-ozone cleaner (NL-UV42, Nippon Laser & Electronics Lab Co. Ltd., Nagoya, Japan) for 10 min to make the glass surface hydrophilic. A polyimide solution (SUNEVER, Nissan Chemical Co., Ltd., Tokyo, Japan) was spin-coated on the glass substrates at 4000 rpm for 40 s with a spin-coater (MS-A100, MIKASA, Co., Ltd., Tokyo, Japan), and heated at 120 °C for 2 h to obtain surface-treated glass substrates, which aligns LC molecules homeotropically (out-of-plane direction). Two surface-treated substrates were bonded with 100-μm-thick polyimide tapes to manufacture a glass cell. The thickness of the glass cell was measured using a micrometer (MDC-25MX, Mitutoyo Co., Kanagawa, Japan).

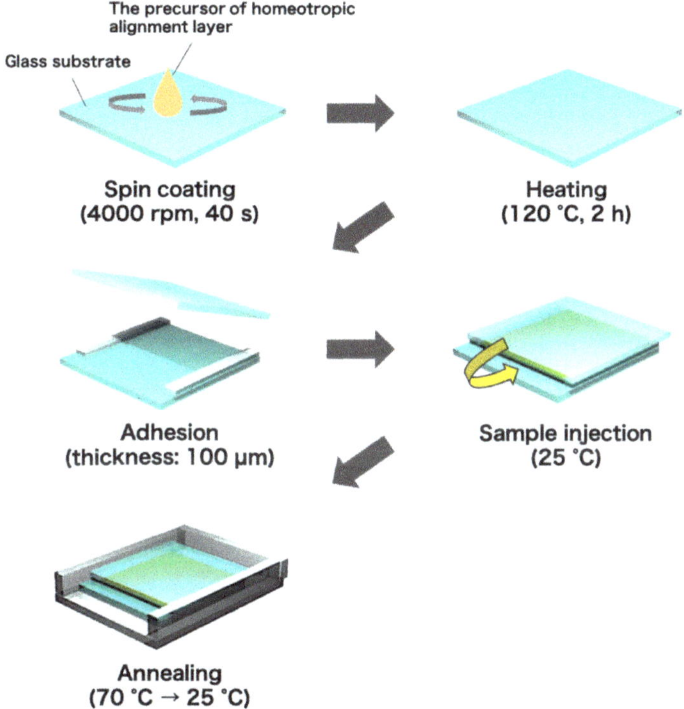

Figure 2. Preparation procedure for homeotropic-aligned oligothiophene-doped LC cells.

The dye-doped host LCs were injected into glass cells by capillary action. The glass cells were heated up to 70 °C to remove the disturbance of the molecular orientation by injection and then cooled down to room temperature at a cooling rate of 2 °C/min. We defined the five prepared cells depending on the compounds as follows: TR5/5CB; TR5/(5CB F-LC1); TR5/(5CB F-LC2); TR5/(5CB F-LC3); and TR5/(5CB F-LC4).

2.2. Evaluation of Initial Molecular Orientation

The initial molecular orientation in the samples was evaluated by conoscopic polarized optical microscopy (POM) and polarized ultraviolet (UV)–visible (vis) absorption spectroscopy. The conoscopic micrographs were obtained using a polarized optical microscope (BX53-P, Olympus Corp., Tokyo, Japan) equipped with an interference filter at

550 nm. The absorption spectra were measured with a UV–vis spectrophotometer (V-750, JASCO Corp., Tokyo, Japan) equipped with a rotatable film holder. Absorbances parallel and perpendicular to the direction of the sample injection were defined as $A_{\parallel}$ and $A_{\perp}$, respectively.

2.3. Self-Diffraction Ring Measurement

Photoinduced molecular reorientation was analyzed from self-diffraction rings. The diffraction ring is formed by the refractive index change of molecular reorientation caused by light irradiation above certain intensity. Irradiation of a homeotropic-aligned cell with a low-intensity Gaussian beam does not cause molecular reorientation (Figure 3a). When the light intensity exceeds the threshold of the molecular reorientation, the diffraction ring occurs due to self-focusing and self-phase modulation (Figure 3b) [8]. The number of rings (N) depends on the photoinduced refractive index change ($\Delta n'$) of the samples, given by

$$N = |\,\Delta n'\,|\,L\lambda^{-1} \tag{1}$$

where L and λ denote the cell thickness and the wavelength of the laser beam, respectively [21]. $\Delta n'$ is the change in the refractive index of the sample. Rod-shaped LC molecules have an anisotropic refractive index in the short and long axes of the molecule. When photoinduced molecular reorientation from perpendicular to parallel to the polarization direction occurs, the refractive index becomes large, resulting in an increase in the number of rings.

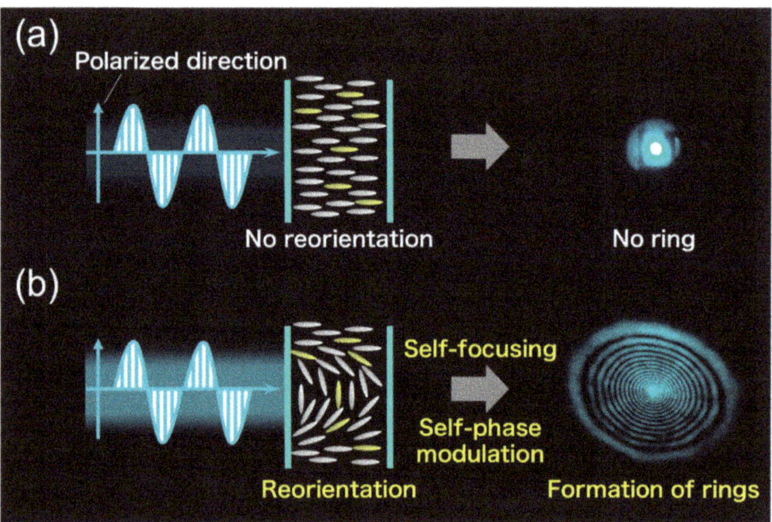

Figure 3. Principle of self-diffraction ring formation: (**a**) below and (**b**) above the threshold light intensity.

The optical setup for self-diffraction measurement is shown in Figure 4. We used a linearly polarized direct diode (DD) laser beam (EXLSR-488C-200-CDRH, Spectra-Physics, Inc., Tokyo, Japan) with a wavelength of 488 nm where TR5 has an absorption band. The beam diameter was expanded from 700 μm to 1.7 mm via lenses L1 and L2. Spatial filtering of the laser beam was performed with a pinhole ($\varphi = 50$ μm) and lenses L3. The laser beam was incident on the sample cell placed at the focal point of L4. The beam diameter at the focal point was 54 μm. The light intensity was controlled using a variable neutral density filter. The light intensity at the irradiation spot was defined as

$$I = I_0 \pi w^2 \tag{2}$$

where w and I_0 are the beam waist and incident light power, respectively. I_0 was measured using a power meter and a beam splitter with a definite ratio (Transmittance:Reflectance = 1:1). Transmitted light was projected onto a screen as self-diffraction rings. As the light intensity increases, the molecular reorientation is further induced, resulting in the increase in the number of rings. We counted the self-diffraction rings manually to evaluate the molecular reorientation of LCs in the cell. The threshold intensity of the molecular reorientation was defined as the light intensity at which the first ring appears. The first ring was monitored with a beam profiler (BGP-USB-SP620, Ophir-Spiricon LLC., North Logan, UT, USA).

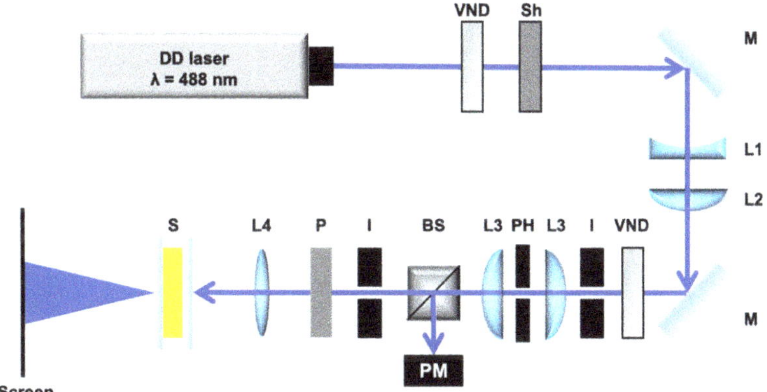

Figure 4. Optical setup for self-diffraction ring measurement. VND, variable neutral density filter; Sh, shutter; M, mirror; L1, plane concave lens ($f = -80$ mm); L2, plane convex lens ($f = 150$ mm); I, iris; L3, plane convex lens ($f = 70$ mm); PH, pinhole; BS, beam splitter; PM, power meter; P, polarizer; L4, biconvex lens ($f = 150$ mm); S, sample.

2.4. Elastic Constant Measurement

A Frank elastic constant of the samples was measured with an elastic constant measurement system (EC-1, TOYO Corp., Tokyo, Japan). Glass cells coated with indium-tin-oxide (ITO) electrodes (KSRP-25/B111P1NSS) required for elastic constant measurement were obtained from EHC Co., Ltd., Tokyo, Japan. The glass substrates (25 mm × 20 mm) with the electrode area (10 mm × 10 mm) were treated with a homogeneous alignment layer (in-plane direction). The cell gap was 25 μm. Each sample was injected into the glass cells by capillary action. The glass cells were heated up to 70 °C and then cooled down to room temperature at a cooling rate of 2 °C/min.

3. Results and Discussion

The prepared samples exhibited an optically transparent yellow color due to the absorption of TR5 (Figure 5a). Conoscopic micrographs of the samples showed a clear isogyre (Figure 5b). This means that the optic axes of LC molecules in the samples are perpendicular to the glass substrate. Furthermore, the UV–vis absorption spectra perpendicular ($A_\perp$) and parallel ($A_\parallel$) to the direction of the sample injection were identical in the wavelength range from 350 to 600 nm (Figure 5c). TR5 is a dichroic dye with anisotropic absorption in the short and long axes of the molecule. Therefore, this result demonstrates that the long axis of the TR5 molecules are perpendicular to both incident orthogonal polarizations and that both host LCs and TR5 in each cell are aligned uniformly and homeotropically.

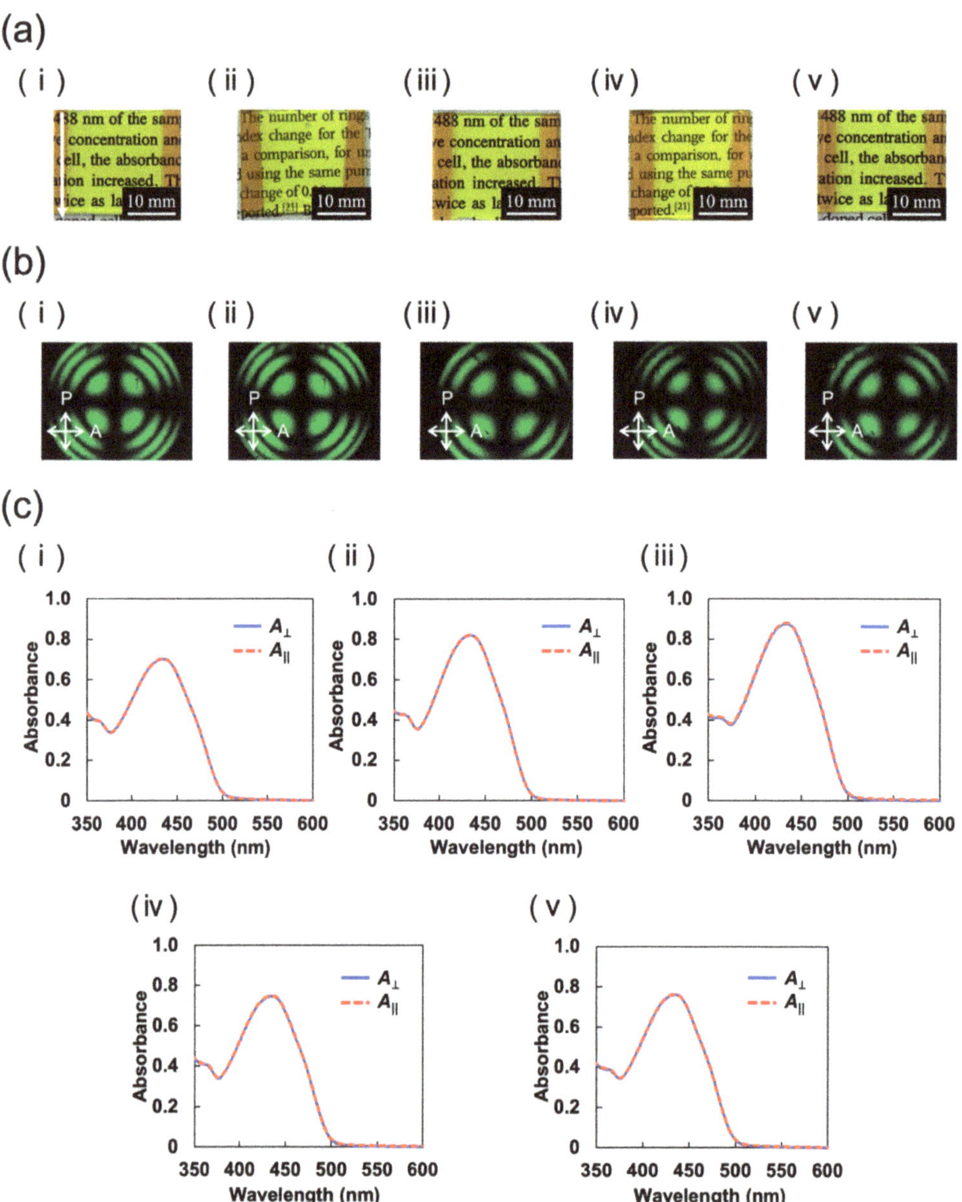

Figure 5. (**a**) Photographs, (**b**) conoscopic POM images, and (**c**) polarized UV-vis absorption spectra of prepared samples: (**i**) TR5/5CB, (**ii**) TR5/(5CB F-LC1), (**iii**) TR5/(5CB F-LC2), (**iv**) TR5/(5CB F-LC3), and (**v**) TR5/(5CB F-LC4). Blue solid and red dash lines in (**c**) denote the absorption perpendicular ($A_{\perp}$) and parallel ($A_{\parallel}$) to the direction of sample injection, respectively.

Table 1 shows the concentration of TR5 in each sample and the absorbance at a wavelength of 488 nm obtained from absorbance spectra of an average of the $A_{\perp}$ and $A_{\parallel}$. An increase in the absorbance at 488 nm decreases the threshold intensity [17]. Therefore, the concentration of TR5 in each sample was adjusted so that the absorbance at 488 nm was almost the same for each sample: the absorbance of TR5/5CB, TR5/(5CB F-LC1), TR5/(5CB F-LC2), TR5/(5CB F-LC3), and TR5/(5CB F-LC4) were 0.155, 0.155, 0.162, 0.161, and 0.158, respectively.

Table 1. The TR5 concentration and absorption at 488 nm of the samples.

Sample	Concentration of TR5 (mol%)	Abs. at 488 nm
TR5/5CB	0.10	0.155
TR5/(5CB F-LC1)	0.13	0.155
TR5/(5CB F-LC2)	0.10	0.162
TR5/(5CB F-LC3)	0.12	0.161
TR5/(5CB F-LC4)	0.10	0.158

The OFT was investigated from the diffraction rings. Figure 6 displays the transmitted light from the TR5/5CB cell. The incidence of a laser beam with a low light intensity did not cause the molecular reorientation, thus no diffraction ring appeared (Figure 6a). When the intensity exceeded 20.1 W/cm^2, the first diffraction ring appeared on the screen (Figure 6b). Furthermore, the number of the rings increased as the light intensity increased up to 54.6 W/cm^2 (Figure 6b–d), and became constant.

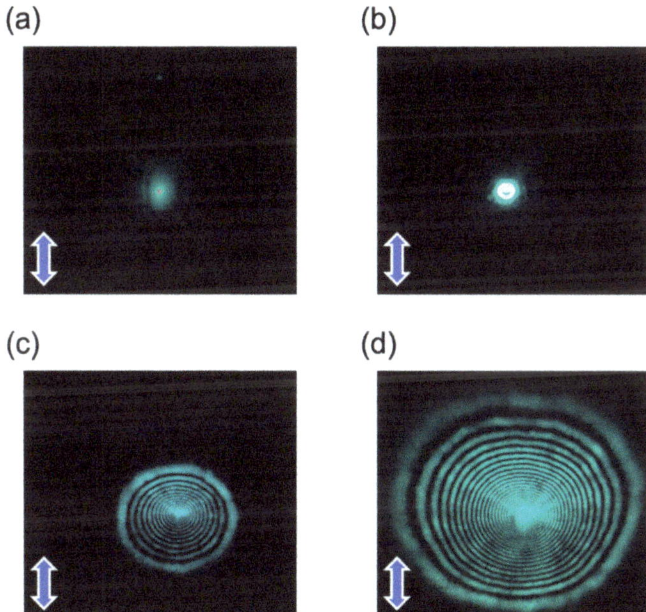

Figure 6. Diffraction ring images of TR5/5CB with the incident light intensity of (**a**) 15, (**b**) 20, (**c**) 27, and (**d**) 55 W/cm^2. Blue arrows indicate the polarization direction of the incident light.

The threshold light intensity inducing the diffraction rings depended on the host LCs. Figure 7 shows the diffraction patterns of each sample formed on a screen at a light intensity of 18 W/cm^2. Diffraction rings appeared in TR5/(5CB F-LC2) and TR5/(5CB F-LC4); in contrast, they did not appear in TR5/5CB, TR5/(5CB F-LC1), and TR5/(5CB F-LC3). This indicates that the sensitivity of light-induced molecular reorientation is different in each sample.

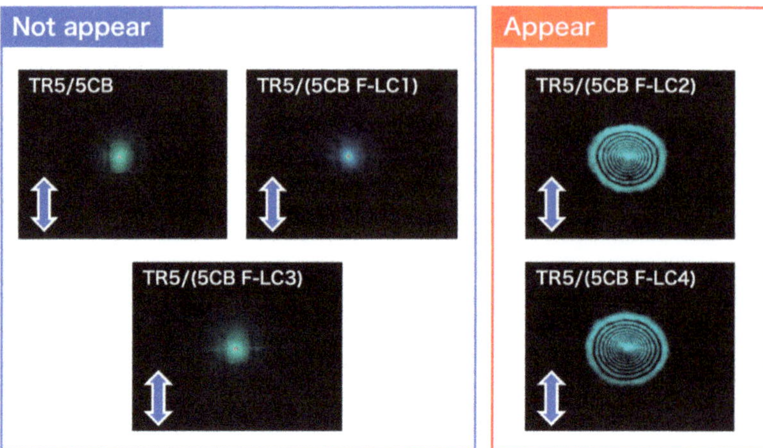

Figure 7. Typical diffraction patterns of all samples induced by a multi-mode laser beam formed at 18 W/cm^2. Blue arrows indicate the polarization direction of the incident light.

Figure 8 shows the number of diffraction rings as a function of light intensity, and Table 2 summarizes the threshold intensity, the maximum number of rings, and K_{33} elastic constant of the samples. The threshold intensities of TR5/(5CB F-LC2) and TR5/(5CB F-LC4), which contain the LCs with three fluorine substituents, were 12.0 W/cm^2 and 11.7 W/cm^2, respectively. These are 42% lower than that of TR5/5CB (20.1 W/cm^2). Furthermore, TR5/(5CB F-LC1) and TR5/(5CB F-LC3), which contain the LCs with two fluorine substituents, exhibited higher threshold intensities than TR5/(5CB F-LC2) and TR5/(5CB F-LC4). This threshold intensity difference indicates that the sensitivity of light-induced molecular reorientation is affected by the number of fluorine substituents in the host LCs. Furthermore, the maximum number of rings of TR5/5CB, TR5/(5CB F-LC1), TR5/(5CB F-LC2), TR5/(5CB F-LC3), and TR5/(5CB F-LC4) were 23, 18, 18, 23, and 22, respectively. TR5/5CB, TR5/(5CB F-LC3), and TR5/(5CB F-LC4), which showed a larger number of diffraction rings, have biphenyl mesogenic structures.

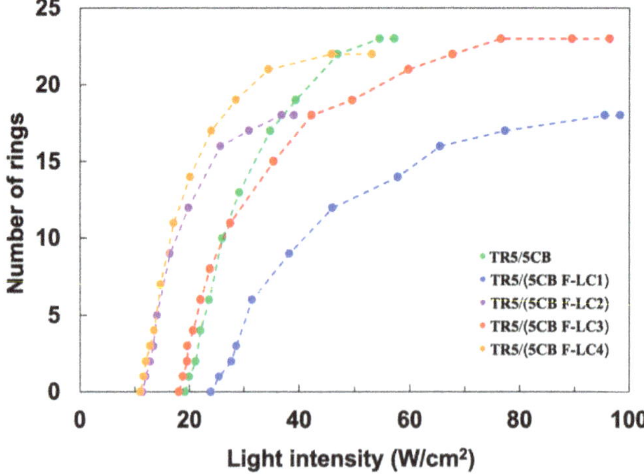

Figure 8. Number of diffraction rings of TR5/5CB, TR5/(5CB F-LC1), TR5/(5CB F-LC2), TR5/(5CB F-LC3),and TR5/(5CB F-LC4) as a function of light intensity.

Table 2. Threshold intensity, the maximum number of rings, and K_{33} elastic constant for TR5/5CB, TR5/(5CB F-LC1), TR5/(5CB F-LC2), TR5/(5CB F-LC3), and TR5/(5CB F-LC4).

Sample	Threshold Intensity (W/cm^2)	Maximum Number of Rings	K_{33} (pN)
TR5/5CB	20.1	23	9.1
TR5/(5CB F-LC1)	25.5	18	11.1
TR5/(5CB F-LC2)	12.0	18	3.1
TR5/(5CB F-LC3)	18.9	23	10.1
TR5/(5CB F-LC4)	11.7	22	6.0

We discuss the OFT of each sample. The number of rings is governed by a sample cell thickness, wavelength of the incident light, and refractive index change, as shown in Equation (1). In this experiment, the cell thickness (100 µm) and the wavelength (488 nm) are the same for all samples. Therefore, the difference in the maximum number of rings is derived from the maximum refractive index change $\Delta n'_{max}$ of the sample. The value of $\Delta n'_{max}$ of 5CB, F-LC1, F-LC2, F-LC3, and F-LC4 was estimated to be 0.11, 0.09, 0.09, 0.11, and 0.11, respectively. 5CB, F-LC3, and F-LC4 with a biphenyl structure have longer conjugation lengths than F-LC1 and F-LC2 with a phenyl ring. The biphenyl structure tends to have a larger birefringence because of the anisotropic delocalization of electrons. This may lead to the difference in the maximum refractive index change of the sample even with the similar molecular reorientation.

The difference in the threshold intensity, which reflects the sensitivity of molecular reorientation, is derived from the dielectric anisotropy of each sample. According to previous work, the threshold intensity is determined by a balance of four torques: optical-electric torque, elastic torque, dye torque, and surface anchoring [15]. The increase in the optical-electric torque promotes the molecular reorientation. The optical-electric torque correlates with the dielectric anisotropy of LCs [10]. In general, the dielectric anisotropy of LCs becomes positive and increases by fluorine substitution [22]. In fact, the three fluorine substituents at the terminal position of a mesogen core enhance the dielectric anisotropy along the long axis of the LC molecule [22,23]. In our materials, the dielectric anisotropy of F-LC2 and F-LC4 modified with three fluorine substituents is larger than that of F-LC1 and F-LC3 modified with two fluorine substituents, which could increase the optical electric torque. Thus, the threshold intensity of TR5/(5CB F-LC2) and TR5/(5CB F-LC4) became lower than that of TR5/(5CB F-LC1) and TR5/(5CB F-LC3).

In addition to the optical-electric torque, the elastic torque of LCs contributes to the sensitivity of the OFT. The elastic torque is proportional to a Frank elastic constant, K_{ii} [11]. In particular, the K_{33} elastic constant of LCs has a large effect on the OFT threshold in a homeotropic cell. Table 2 shows the K_{33} elastic constant of each sample. The values of K_{33} elastic constant of TR5/5CB, TR5/(5CB F-LC1), TR5/(5CB F-LC2), TR5/(5CB F-LC3), and TR5/(5CB F-LC4) were 9.1, 11.1, 3.1, 10.1, and 6.0, respectively. This result indicates that a lower K_{33} elastic constant enhances the sensitivity of the OFT.

Furthermore, we assume that the dye torque was changed by the substituents of host LCs. Marrucci reported that the intermolecular interaction between the anthraquinone dye and the host LC changes according to the substituents of the dye, leading to a change in the dye torque [10,24,25]. Although the mechanism of enhancing the OFT sensitivity in oligothiophene and anthraquinone systems may be different, intermolecular interactions between the oligothiophene dye and host LCs could vary depending on the terminal substituents of the host LCs, resulting in the dye torque change. This study focused on the optical-electric torque, the elastic torque, and the dye torque, while the effect of the optical-electric torque and the dye torque on the OFT sensitivity is not identified experimentally. Surface anchoring may also affect the OFT sensitivity. Detailed analysis is under investigation. The experimental identification of the mechanism will aid in the development of high-performance optical devices.

4. Conclusions

In conclusion, we investigated the OFT of oligothiophene-doped LCs using various host LCs with fluorine substituents. The molecular reorientation was induced by the irradiation with a linearly polarized laser beam. We found that the maximum number of rings of TR5/5CB, TR5/(5CB F-LC3), and TR5/(5CB F-LC4) with biphenyl structures was more than TR5/(5CB F-LC1) and TR5/(5CB F-LC2) with a phenyl ring. The difference in the conjugation length depending on the number of biphenyl rings affected the maximum number of rings. Furthermore, the threshold intensities of TR5/(5CB F-LC2) and TR5/(5CB F-LC4) became 42% lower than that of TR5/5CB. The reduction of the threshold can be explained by the changes in the optical-electric torque, the elastic torque, and the dye torque. The study revealed that the structure of the host LCs affected the sensitivity of the OFT.

Author Contributions: Conceptualization, A.S.; methodology, J.Y., K.M. and K.U.; investigation, J.Y., K.M., K.U. and S.K.; visualization, J.Y., K.M. and K.U.; writing—original draft preparation, J.Y.; writing—review and editing, S.K. and A.S. All authors have read and agreed to the published version of the manuscript.

Funding: This research was supported by a Grant-in-Aid for Scientific Research on Innovative Areas "Molecular Engine" (JSPS KAKENHI Grant Number JP18H05422) and JST CREST Grant Number JPMJCR1814. This work was performed under the Cooperative Research Program of "Network Joint Research Center for Materials and Devices". This work was performed under the Research Program of "Dynamic Alliance for Open Innovation Bridging Human, Environment and Materials" in "Network Joint Research Center for Materials and Devices".

Data Availability Statement: The authors confirm that the data supporting the findings of this study are available within the article.

Acknowledgments: The authors thank Yoshiyuki Kaneko (TOYO Corp., Tokyo, Japan) for his assistance in measuring the elastic constant. The authors also thank Masayuki Kishino (Tokyo Institute of Technology) for his discussions and assistance. The authors also thank Carlos Mejia (Tokyo Institute of Technology) for editing this manuscript.

Conflicts of Interest: The authors declare no conflict of interest.

References

1. Schadt, M. Milestone in the history of field-effect liquid crystal displays and materials. *Jpn. J. Appl. Phys.* **2009**, *48*, 03B001. [CrossRef]
2. Lampert, C.M. Large-area smart glass and integrated photovoltaics. *Sol. Energy Mater. Sol. Cells* **2003**, *76*, 489–499. [CrossRef]
3. Priimagi, A.; Barrett, C.; Shishido, A. Recent twists in photoactuation and photoalignment control. *J. Mater. Chem. C* **2014**, *2*, 7155–7162. [CrossRef]
4. Tabiryan, N.V.; Sukhov, A.V.; Zel'Dovich, B.Y. Orientational optical nonlinearity of liquid crystals. *Mol. Cryst. Liq. Cryst.* **1986**, *136*, 1–139. [CrossRef]
5. Simoni, F.; Bartolino, R. Nonlinear Optical Behavior of Hybrid Aligned Nematic Liquid Crystals. *Opt. Commun.* **1985**, *53*, 210–212. [CrossRef]
6. Zel'dovich, B.Y.; Pilipetskii, N.F.; Sukhov, A.V.; Tabiryan, N.V. Giant optical nonlinearity in the mesophase of a nematic liquid crystal (NCL). *JETP Lett.* **1980**, *31*, 263–269.
7. Zolot'ko, A.S.; Kitaeva, V.F.; Kroo, N.; Sobolev, N.N.; Chilag, L. Finding of optical nonlinearity. *JETP Lett.* **1980**, *32*, 158–162.
8. Durbin, S.D.; Arakelian, S.M.; Shen, Y.R. Optical-field-induced birefringence and Freedericksz transition in a nematic liquid crystal. *Phys. Rev. Lett.* **1981**, *47*, 1411–1414. [CrossRef]
9. Janossy, I.; Lloyd, A.D.; Wherrett, B.S. Anomalous Optical Freedericksz Transition in an Absorbing Liquid Crystal. *Mol. Cryst. Liq. Cryst.* **1990**, *179*, 1–12. [CrossRef]
10. Marrucci, L. Mechanisms of giant optical nonlinearity in light-absorbing liquid crystals: A brief primer. *Liq. Cryst. Today* **2002**, *11*, 6–33. [CrossRef]
11. Janossy, I.; Lloyd, A.D. Low-Power Optical Reorientation in Dyed Nematics. *Mol. Cryst. Liq. Cryst.* **1991**, *203*, 77–84. [CrossRef]
12. Zhang, H.; Shiino, S.; Shishido, A.; Kanazawa, A.; Tsutsumi, O.; Shiono, T.; Ikeda, T. A thiophene liquid crystal as a novel π-Conjugated dye for photo-manipulation of molecular alignment. *Adv. Mater.* **2000**, *12*, 1336–1339. [CrossRef]
13. Matsumoto, K.; Usui, K.; Akamatsu, N.; Shishido, A. Molecular reorientation behavior of oligothiophene-doped polymer-stabilized liquid crystals irradiated with collimated laser beam. *Mol. Cryst. Liq. Cryst.* **2020**, *713*, 46–54. [CrossRef]

14. Usui, K.; Matsumoto, K.; Katayama, E.; Akamatsu, N.; Shishido, A. A Deformable Low-Threshold Optical Limiter with Oligothiophene-Doped Liquid Crystals. *ACS Appl. Mater. Interfaces* **2021**, *13*, 23049–23056. [CrossRef]
15. Usui, K.; Katayama, E.; Wang, J.; Hisano, K.; Akamatsu, N.; Shishido, A. Effect of surface treatment on molecular reorientation of polymer-stabilized liquid crystals doped with oligothiophene. *Polym. J.* **2017**, *49*, 209–214. [CrossRef]
16. Wang, J.; Aihara, Y.; Kinoshita, M.; Mamiya, J.I.; Priimagi, A.; Shishido, A. Laser-Pointer-Induced Self-Focusing Effect in Hybrid-Aligned Dye-Doped Liquid Crystals. *Sci. Rep.* **2015**, *5*, 9890. [CrossRef]
17. Yaegashi, M.; Shishido, A.; Shiono, T.; Ikeda, T. Effect of ester moieties in dye structures on photoinduced reorientation of dye-doped liquid crystals. *Chem. Mater.* **2005**, *17*, 4304–4309. [CrossRef]
18. Yaegashi, M.; Kinoshita, M.; Shishido, A.; Ikeda, T. Direct Fabrication of Microlens Arrays with Polarization Selectivity. *Adv. Mater.* **2007**, *19*, 801–804. [CrossRef]
19. Aihara, Y.; Kinoshita, M.; Wang, J.; Mamiya, J.I.; Priimagi, A.; Shishido, A. Polymer stabilization enhances the orientational optical nonlinearity of oligothiophene-doped nematic liquid crystals. *Adv. Opt. Mater.* **2013**, *1*, 787–791. [CrossRef]
20. Wang, J.; Aihara, Y.; Kinoshita, M.; Shishido, A. Effect of polymer concentration on self-focusing effect in oligothiophene-doped polymer-stabilized liquid crystals. *Opt. Mater. Exp.* **2015**, *5*, 538–548. [CrossRef]
21. Durbin, S.D.; Arakelian, S.M.; Shen, Y.R. Laser-induced diffraction rings from a nematic-liquid-crystal film. *Opt. Lett.* **1981**, *6*, 411–413. [CrossRef] [PubMed]
22. Hird, M. Fluorinated liquid crystals—Properties and applications. *Chem. Soc. Rev.* **2007**, *36*, 2070–2095. [CrossRef] [PubMed]
23. Zhu, S.; Chen, R.; Zhang, W.; Niu, X.; Chen, W.; Mo, L.; Hu, M.; Zhang, L.; Li, J.; Chen, X.; et al. Dissecting terminal fluorinated regulator of liquid crystals for fine-tuning intermolecular interaction and molecular configuration. *J. Mol. Liq.* **2020**, *310*, 113225. [CrossRef]
24. Marrucci, L.; Paparo, D.; Maddalena, P.; Massera, E.; Prudnikova, E.; Santamato, E. Role of guest-host intermolecular forces in photoinduced reorientation of dyed liquid crystals. *J. Chem. Phys.* **1997**, *107*, 9783–9793. [CrossRef]
25. Marrucci, L.; Paparo, D.; Vetrano, M.R.; Colicchio, M.; Santamato, E.; Viscardi, G. Role of dye structure in photoinduced reorientation of dye-doped liquid crystals. *J. Chem. Phys.* **2000**, *113*, 10361–10366. [CrossRef]

Article

Size-Controllable Synthesis of Monodisperse Magnetite Microparticles Leading to Magnetically Tunable Colloidal Crystals

Toya Seki, Yutaro Seki, Naoto Iwata and Seiichi Furumi *

Department of Chemistry, Graduate School of Science, Tokyo University of Science, 1-3 Kagurazaka, Shinjuku, Tokyo 162-8601, Japan; 1322578@ed.tus.ac.jp (T.S.); 1320568@alumni.tus.ac.jp (Y.S.); n-iwata@rs.tus.ac.jp (N.I.)
* Correspondence: furumi@rs.tus.ac.jp; Tel.: +81-3-3260-4271

Abstract: Colloidal crystals (CCs) are periodic arrays of monodisperse microparticles. Such CCs are very attractive as they can be potentially applicable as versatile photonic devices such as reflective displays, sensors, lasers, and so forth. In this article, we describe a promising methodology for synthesizing monodisperse magnetite microparticles whose diameters are controllable in the range of 100–200 nm only by adjusting the base concentration of the reaction solution. Moreover, monodisperse magnetite microparticles in aqueous suspensions spontaneously form the CC structures under an external magnetic field, leading to the appearance of Bragg reflection colors. The reflection peak can be blue-shifted from 730 nm to 570 nm by the increase in the external magnetic field from 28 mT to 220 mT. Moreover, the reflection properties of CCs in suspension depend on the microparticle concentration in suspension and the diameter of the magnetite microparticles. Both fine-control of microparticle diameter and investigation of magneto-optical properties of CCs would contribute to the technological developments in full-color reflective displays and sensors by utilizing these monodisperse magnetite microparticles.

Keywords: colloidal crystals; magnetite; microparticles; Bragg reflection; magnetic response

Citation: Seki, T.; Seki, Y.; Iwata, N.; Furumi, S. Size-Controllable Synthesis of Monodisperse Magnetite Microparticles Leading to Magnetically Tunable Colloidal Crystals. *Materials* **2022**, *15*, 4943. https://doi.org/10.3390/ma15144943

Academic Editors: Ivan V. Timofeev and Wei Lee

Received: 7 June 2022
Accepted: 29 June 2022
Published: 15 July 2022

Publisher's Note: MDPI stays neutral with regard to jurisdictional claims in published maps and institutional affiliations.

1. Introduction

Colloidal crystals (CCs) are periodic arrays of monodisperse microparticles [1–5]. Such CCs can be regarded as one of the three-dimensional (3D) photonic crystals that show the forbidden regions for photons in the dispersion spectrum, that is, the photonic bandgaps (PBGs), due to the spatial periodicity of the refractive indices between the colloidal microparticles and dispersion media [6–8]. One of the most important properties of CCs is their ability to show visible reflection using colloidal microparticles with diameters of several hundred nanometers. The reflection peak wavelength (λ) of CCs is approximately calculated according to the following Bragg's equation [8,9].

$$\lambda = 2nd \sin \theta \tag{1}$$

where n is the effective refractive index of materials, d is the interparticle spacing of CC, and θ is the angle of incident light, that is, the Bragg angle. When the value of λ is comparable to the wavelength range of visible light, the PBGs of CCs can be observed as Bragg reflection colors. Introduction of stimuli-responsive materials in the microparticles and/or the background encompassing them ensures the on-demand control of reflection peaks of CCs by external stimuli such as temperature, ionic strength, mechanical force, and so forth [10–13]. Owing to this unique optical property, the CCs are very attractive as they can be potentially applicable in versatile photonic devices as color reflective displays, sensors, lasers, and so forth.

Previously, some types of methods for preparing monodisperse magnetite microparticles have been reported [14]. These microparticles can be applied to CCs, which can be categorized into two types: 3D or one-dimensional (1D) CCs [15–19]. For instance, Asher and co-workers reported that highly charged monodisperse polystyrene (PS) microparticles encapsulating superparamagnetic nanoparticles can self-assemble into the 3D CC structures due to their highly charged surface [19]. The attractive force between microparticles induced by applying an external magnetic field was not strong because of the low content of the magnetic component in the microparticles, leading to the limited tuning range of Bragg reflection wavelength and insufficient magnetic responsivity of the CCs. On the other hand, Yin and co-workers developed an intriguing strategy for synthesizing polyelectrolyte-capped monodisperse superparamagnetic magnetite microparticles that can be directly used as the building blocks of the 1D CC structures [17]. These magnetite microparticles are advantageous in the following three aspects. First, they can show higher saturated magnetizations and superior magnetic responsivity because they only consist of magnetite. Second, the magnetite microparticles can be easily dispersed in water due to their highly negatively charged surface offered by polyelectrolyte surfactants. Finally, magnetite microparticle diameters can be controlled in the range between 30 nm and 180 nm with narrow size distribution. By utilizing these characteristics, the magnetite microparticles in aqueous suspensions spontaneously form the 1D chain-like CC structures under an external magnetic field, leading to the appearance of Bragg reflection colors due to the spatial modulation of the refractive indices between magnetite microparticles and water as a dispersion medium. It has been considered that the 1D chain-like CC structure can be formed by the balance of attractive force induced by magnetic dipole–dipole interactions and electrostatic repulsive force between the microparticles [14]. This means that the interparticle spacing, corresponding to d in Equation (1), can be controlled by changing the attractive force, accompanied by the increase in magnetic field strength. Thus, the Bragg reflection wavelength of the CCs can be easily tuned by the external magnetic field strength. Although there have been numerous precedents with regard to the CC structures of magnetite microparticles, the 1D CC structures are more likely assembled by applying a magnetic field according to the review by Chen [14]. However, the synthetic method of monodisperse magnetite microparticles is very limited. In addition, there have been few reports on the control of microparticle diameter. Considering the possibility of applying these magnetite microparticles to magnetically responsive photonic devices, it is highly demanded that a procedure to synthesize them with desired microparticle diameters is found.

In this article, we report on a promising methodology to synthesize monodisperse magnetite microparticles with diameters of 100–200 nm by adjusting the base concentration of the reaction solution. This size-controllable synthesis procedure is very simple. These magnetite microparticles self-organize the CC structures with visible reflection, even in aqueous suspensions, by applying external magnetic field. The reflection peak wavelength of the CCs can be tuned by the external magnetic field strength, microparticle diameter, and microparticle concentration. Such tunability of magnetically responsive CCs would contribute to the technological developments of next-generation photonic devices such as full-color reflective displays and sensors.

2. Experimental Section

Anhydrous ferric chloride ($FeCl_3$) was obtained from Nacalai Tesque (Kyoto, Japan). Ethylene glycol, anhydrous sodium acetate, and sodium hydroxide (NaOH) were purchased from FUJIFILM Wako Pure Chemical Corporation (Osaka, Japan). Poly(4-stylenesulfonic acid-*co*-maleic acid) sodium salt (PSSMA) with weight average molecular weight of ~2.0×10^4 was obtained from Sigma-Aldrich Japan (Tokyo, Japan). The molar ratio of poly(4-styrenesulfonic acid) unit and poly(maleic acid) unit was 1:1, according to the data sheet of the manufacturer.

Monodisperse magnetite microparticles were prepared by one-pot hydrothermal synthesis using $FeCl_3$ as a precursor. In this study, NaOH concentration was varied in the range of 0.28–0.38 M, as shown in Table 1. Here, a typical synthesis procedure of the monodisperse magnetite microparticles at an NaOH concentration of 0.28 M is described as follows (Table 1, sample code: **M-1**).

Table 1. Amounts and concentrations of NaOH in the hydrothermal synthesis and size properties and saturation magnetization (M_s) values of the resultant magnetite microparticles.

Sample	Amount of NaOH (g)	NaOH Concentration (M)	Reaction Temperature (°C)	Diameter (nm) [1]	CV (%) [1]	D [2] (nm)	M_s [3] (emu/g)
M-1	0.36	0.28	190	106	10.8	7.3	58.1
M-2	0.40	0.31	190	111	10.4	24	60.5
M-3	0.44	0.34	190	150	6.4	37	63.0
M-4	0.48	0.38	190	200	7.0	44	82.3
M-5	0.48	0.38	210	226	25.8	− [4]	− [4]
M-6	0.48	0.38	230	226	18.6	− [4]	− [4]

[1] Determined by SEM observation. [2] Determined by Debye–Scherrer equation using the (311) diffraction peak in the XRD pattern. [3] Determined by SQUID measurement. [4] Not measured.

A mixture of $FeCl_3$ (0.52 g, 3.2 mmol), ethylene glycol (32 mL, 0.58 mol), anhydrous sodium acetate (2.4 g, 30 mmol), PSSMA (0.80 g, 4.5 mmol, monomer equivalent), and 120 μL of ultrapure water was vigorously stirred for 20 min at room temperature. After that, NaOH (0.36 g, 12 mmol) was added to this mixture and stirred at room temperature until completely dissolved. The resulting mixture was transferred into a 50 mL Teflon-lined stainless-steel autoclave and heated at 190 °C for 16 h. After cooling down to room temperature, the products were collected by a handheld magnet and washed three times with a mixture of water and ethanol at a volume ratio of 2:1, and finally dried in a vacuum for 24 h.

X-ray diffraction (XRD) patterns of magnetite microparticles were acquired by an X-ray diffractometer (RINT2500, Rigaku, Tokyo, Japan) using Cu Kα radiation under the acceleration voltage of 40 kV. Morphological size and shape of as-prepared magnetite microparticles were characterized by a field-emission scanning electron microscopy (SEM; JSM-7800F Prime, JEOL, Tokyo, Japan) operated at an acceleration voltage of 15 kV. The samples for SEM observation were prepared by depositing the dilute suspensions of magnetite microparticles on a silicon wafer and subsequent drying them at room temperature. Before SEM observation, the sample surface was coated with a thin layer of osmium oxide. Coefficients of variation (CV) in microparticle diameter, which are numerically defined as the ratio of the standard deviation to the average diameter, were calculated from the SEM images. Magnetic properties of magnetite microparticles at 27 °C were measured by using a superconducting quantum interference device (SQUID) magnetometer (MPMS-XL7AC, Quantum Design, San Diego, CA, USA). Zeta potential measurements were conducted using a microparticle charge analysis system (Zetasizer Nano ZS, Malvern Panalytical Ltd., Worcestershire, UK).

Reflection spectra of aqueous suspensions of magnetite microparticles were taken on a compact charge-coupled device spectrometer (USB 2000, Ocean Optics, Orlando, FL, USA) equipped with a halogen light source (HL-2000, Ocean Optics) and an optical fiber with a reflection probe (R200-7-UV-VIS, Ocean Optics). After the aqueous suspension of magnetite microparticles was placed in a quartz glass cuvette with an optical path length of 1.0 cm, the reflection spectra were measured upon applying the external magnetic field, which was generated by an NdFeB magnet with a size of 4 × 4 × 1 cm (NK066, Niroku Seisakusho, Kobe, Japan). The magnetic field strength was tuned by adjusting the distance between the cuvette and the NdFeB magnet and measured using a tesla meter (TM-801, KANETEC Co., Ltd., Nagano, Japan).

3. Results and Discussion

3.1. XRD Measurements of Magnetite Microparticles

As-prepared colloidal microparticles exhibited intrinsic color of dark brown, arising from light absorption of magnetite. Therefore, we measured XRD patterns of the microparticles in order to characterize their chemical identity. Figure 1 shows XRD patterns of a series of microparticles of **M-1**–**M-4** synthesized at different NaOH concentrations in the hydrothermal synthesis. All the peak positions in each XRD pattern were almost identical, regardless of the difference in synthetic conditions of magnetite microparticles. It is plausible that the peaks at 2θ of 30.0°, 35.3°, 36.9°, 42.9°, 53.2°, 56.7°, and 62.3° are assigned to the diffractions of (220), (311), (222), (400), (422), (511), and (440) lattice planes of the face-centered cubic lattices of magnetite, respectively. In addition, these XRD peaks can be indexed to the JCPDS file, No. 19-0629. Therefore, from these XRD results, we concluded that the colloidal microparticles consisting of magnetite can be prepared by the hydrothermal synthesis of $FeCl_3$ at NaOH concentrations of 0.28–0.38 M.

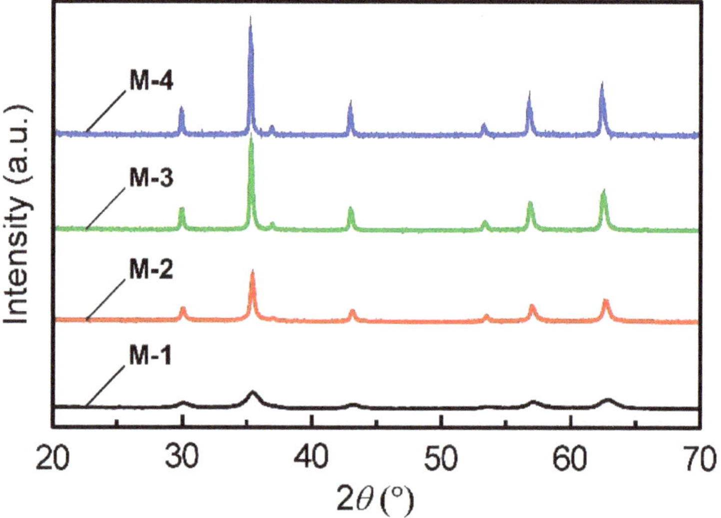

Figure 1. XRD patterns of the magnetite microparticles of **M1**–**M4** with different diameters prepared by the hydrothermal synthesis.

3.2. Size Control of Magnetite Microparticles

Morphological size and shape of the magnetite microparticles were analyzed by SEM observation. The SEM image of **M-1** is shown in Figure 2A. The average diameter and the CV value were determined to be 106 nm and 10.8%, respectively. In general, the monodispersity of microparticle diameter is indispensable to the formation of well-ordered CC structures. For instance, it has been reported that the CV values of colloidal microparticles used for the fabrication of CCs were 5.5% in the case of silica microparticles and 4.2% for PS microparticles in the previous studies [20,21]. By considering the empirical facts, we anticipated that **M-1** can be used for the preparation of CCs owing to its moderately low CV value. However, the detailed SEM observation revealed that the surface of **M-1** is not smooth, in contrast to the conventional colloidal microparticles of silica or PS. This morphological result infers that **M-1** consists of spherically shaped clusters of small magnetite particles with a diameter of several nanometers [22]. The mechanism for the formation of monodisperse magnetite clusters can be explained as follows. The growth process of magnetite microparticles can be divided into two sequential steps according to the previous work by Yu and co-workers [23]. The first step is nucleation, in which primary magnetite nanocrystals (NCs) are formed by the hydrolysis of $FeCl_3$. Subsequently, these magnetite NCs aggregate into clusters to reduce their surface energy. Because this growth process of magnetite microparticles, that is, the aggregation of magnetite NCs, selectively occurs inside the PSSMA network, magnetite microparticles with narrow size distribution can be synthesized [24]. From these results, the magnetite microparticles exhibited reasonable monodispersity by the hydrothermal synthesis of $FeCl_3$ under this reaction condition.

Figure 2. (A–F) SEM images of a series of magnetite microparticles of **M-1–M-6**. All white scale bars represent 100 nm.

Subsequently, we attempted to control the magnetite microparticle diameter by changing the reaction conditions. This is because the control of microparticle diameter is of paramount importance from both scientific and technological perspectives of CC structures. Because the Bragg reflection wavelength of CCs is dependent on the lattice constant, which is affected by the difference in microparticle diameter, in this work, another series of magnetite microparticles were synthesized at the different NaOH concentrations from 0.31 M to 0.38 M. These magnetite microparticles, noted as **M-2**, **M-3**, and **M-4** in Table 1, were spherically-shaped clusters of small magnetite microparticles similar to **M-1**, as confirmed from SEM images (Figure 2B–D). Although **M-1** showed 106 nm in the microparticle diameter of, as mentioned above, the average diameters of **M-2–M-4**, these increased from 111 nm to 200 nm, accompanied by an increase of NaOH concentrations. However, the CV values were approximately 6–10%, regardless of the NaOH concentrations (Table 1). Therefore, **M-2–M-4** were also found to be magnetite microparticles with relative monodispersity. When the NaOH concentration was further increased to 0.44 M, magnetite microparticles also showed ~220 nm in diameter. However, these magnetite microparticles have poor dispersibility in ultrapure water, probably due to the sedimentation caused by the increase in microparticle weights or by the enhancement of van der Waals attractive force induced by the increase in particle diameter. The increase in magnetite microparticle diameter can be explained by the increase in the size of primary NCs. Here, the size of primary NCs (D) was estimated by the following Debye–Scherrer equation:

$$D = \frac{0.9 \times \lambda}{\beta_{1/2} \times \cos\theta} \tag{2}$$

where λ stands for the wavelength of Cu Kα radiation, corresponding to 0.154 nm, θ is the diffraction angle, and $\beta_{1/2}$ is the full width at half maximum value of a diffraction peak. In this work, we focused on an outstanding diffraction peak of (311) lattice planes at 2θ of 35.3°. By applying the $\beta_{1/2}$ values of (311) diffraction peaks in the XRD patterns into Equation (2), the D values of **M-1**, **M-2**, **M-3**, and **M-4** were estimated to be 7.3 nm, 24 nm, 37 nm, and 44 nm, respectively (Table 1). In addition, the diffraction peaks were also intensified owing to the increase in the size of primary NCs. Thus, it can be anticipated that the increase in magnetite microparticle diameter is triggered by the increase in the size of primary NCs. We also considered that the higher alkaline conditions accelerate the hydrolysis of FeCl$_3$, which result in the formation of larger primary magnetite NCs.

Finally, we tried to further increase the microparticle diameter by elevating the reaction temperature. Figure 2E,F show the SEM images of **M-5** and **M-6**, which were prepared by the hydrothermal synthesis conducted at 210 °C and 230 °C, respectively. **M-5** and **M-6** were also found to be magnetite microparticles, as confirmed from XRD measurements (Supplementary Materials, Figure S1). However, the diameters of each magnetite microparticle were almost unchanged when compared to **M-4**. Moreover, as evident from SEM images, the CV values of **M-5** and **M-6** deteriorated to around 20%, probably because the nucleate growth of magnetite NCs was uncontrollable. From these results, we found that the appropriate temperature for the synthesis of monodisperse magnetite microparticles is 190 °C and the microparticle diameter can be controlled by changing the NaOH concentration from 0.28 M to 0.38 M.

3.3. Water Dispersibility and Magnetic Properties of Magnetite Microparticles

The magnetite microparticles of **M-1**–**M-4** could be easily dispersed in ultrapure water. This was achieved by the removal of excess ions by washing with a mixture of ethanol and water as well as the surface modification of magnetite microparticles with PSSMA. As mentioned in the Experimental Section, PSSMA is a copolymer of poly(*p*-styrenesulfonic acid) and poly(maleic acid). The sulfonic groups in poly(*p*-styrenesulfonic acid) moieties are completely ionized in aqueous solutions while the carboxy groups in poly(maleic acid) moieties can strongly coordinate with iron cations on the magnetite microparticle surface. Thus, the outermost surface of magnetite microparticles is highly negatively charged to offer high dispersibility in water. To evaluate the dispersibility of magnetite microparticles, we measured the zeta potential of **M-3**. As a result, the zeta potential of **M-3** was determined to be −45 mV. This value of zeta potential is comparable to that of monodisperse silica microparticles used for the fabrication of CCs in a previous study [25]. Thus, it can be anticipated that the magnetite microparticles hardly aggregate in water owing to the large electrostatic repulsive force offered by the highly negatively charged surface. By considering these facts, we concluded that the magnetite microparticles are suitable for the fabrication of CCs owing to their monodispersity and high dispersibility in water.

The magnetic properties of magnetite microparticles of **M-1**–**M-4** were analyzed by SQUID measurements. Figure 3 shows the magnetic hysteresis loops of **M-1**–**M-4** measured at 27 °C. The magnetite microparticles showed no remanence and coercivity, indicative of their superparamagnetic feature. As the external magnetic field increases, the induced magnetization saturates at a certain value, which is so-called saturation magnetization (M_s). From the measurements, the M_s values of **M-1**, **M-2**, **M-3**, and **M-4** were 58.1 emu/g, 60.5 emu/g, 63.0 emu/g, and 82.3 emu/g, respectively. The M_s values were monotonously increased in the order of microparticle diameter. The increase in M_s value can be explained by the large magnetic domains caused by the increase in microparticle diameter. Previously, it has been reported that the M_s value of polystyrene encapsulating superparamagnetic microparticles used for the fabrication of CCs was 38.6 emu/g, which is much lower than those of **M-1**–**M-4** [26]. Thus, it can be anticipated that these magnetite microparticles of **M-1**–**M-4** can show superior magnetic responsivity when they are applied to CCs.

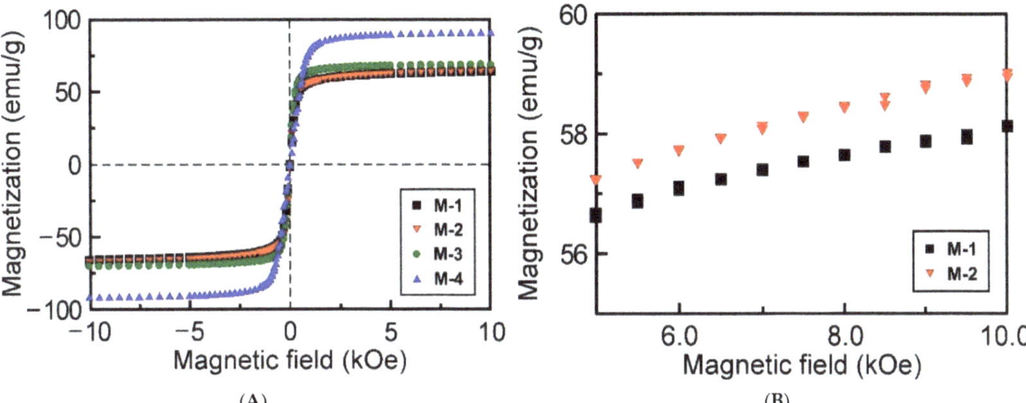

(A) (B)

Figure 3. (**A**) Magnetic hysteresis loops of magnetite microparticles measured at 27 °C. (**B**) An enlarged profiles of **M-1** and **M-2** in the range of 5 kOe to 10 kOe for comparison of their magnetization difference.

3.4. Optical Properties of Magnetite Microparticle Suspensions

Under an external magnetic field, the magnetite microparticles were quickly self-assembled into CC arrays in fluid aqueous suspensions owing to their highly charged surface, resulting in the appearance of visible Bragg reflection. Monodisperse magnetite microparticles in suspensions are known to spontaneously form the 1D chain-like CC structures under an external magnetic field, which can be formed by the balancing of the interaction of magnetic dipole–dipole attractions and electrostatic repulsive force between microparticles [14]. As the force balance fluctuates by tuning the magnetic field strength, the Bragg reflection wavelength shifts immediately triggered by changes in the spacing between the lattice plane of the 1D chain-like CC structures.

Figure 4A,B show the changes in snapshot image and the reflection spectrum of an aqueous suspension of **M-3** at the microparticle concentration of 0.30 wt%, respectively, upon changing the external magnetic field in a continuous way. The Bragg reflection colors were observed only along the direction of external magnetic field, implying that magnetite microparticles in suspensions form the 1D chain-like CC structures under an external magnetic field. The reflection spectra were measured for magnetite microparticle suspensions placed in a quartz glass cuvette. The external magnetic field was generated by an NdFeB magnet, and its strength was controlled by adjusting the distance between the NdFeB magnet and the cuvette. The reflection color was changed from red to green by moving the magnet at a distance from 4.0 cm to 0.5 cm (Supplementary Materials, Figure S2), corresponding to the external magnetic field from 28 mT to 220 mT, respectively (Figure 4A).

We also observed that the Bragg reflection peak blue-shifts from 730 nm to 570 nm in a continuous way upon increasing the magnetic field (Figure 4B). Such a blue-shift of reflection peak wavelength can be explained as follows. Applying the aqueous suspension of **M-3** at relatively high external magnetic field brings about enhancement in attractive forces between the magnetite microparticles, thereby leading to the geometric decrease in interparticle spacing of the 1D chain-like CC structure. As a result, the Bragg reflection wavelength shifts to shorter wavelengths as the interparticle spacing decreases. Thus, the Bragg reflection wavelength of the 1D chain-like CC structure was found to depend on the external magnetic field strength. Notably, the magnetic response of the Bragg reflection color changes of magnetite microparticle suspension was very quick and fully reversible, as can be seen in the demonstration video (Supplementary Materials, Video S1), which enabled the on-demand tuning of the Bragg reflection peak of the CC by applying with the magnetic field.

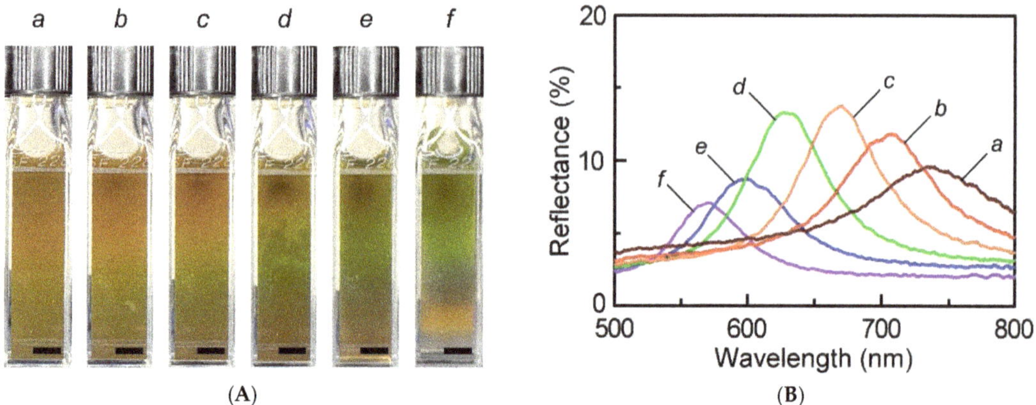

Figure 4. Changes in snapshot image (**A**) and reflection spectrum (**B**) of a suspension of M-3 as a function of the external magnetic field of 28 mT (a), 36 mT (b), 48 mT (c), 65 mT (d), 90.4 mT (e), and 220 mT (f). All black scale bars in the images represent 0.5 cm.

According to another finding, the Bragg reflection wavelength was also affected by the microparticle concentrations. To demonstrate this ability, three different kinds of suspensions of **M-3** were prepared at the concentrations of 0.30 wt%, 0.45 wt%, and 0.60 wt%. The dependences of the Bragg reflection wavelength on the microparticle concentration are shown in Figure 5A. By comparing the Bragg reflection wavelength under the same magnetic field strength in the range from 36 mT to 125 mT, it turned out that the Bragg reflection wavelength shifts to a shorter wavelength as the microparticle concentration increases. For example, when an external magnetic field with 125 mT was applied to each suspension, the Bragg reflection wavelength shifted from 473 nm to 582 nm by diluting the microparticle concentration from 0.60 wt% to 0.30 wt%. This happened from the geometric decrease in lattice constant between the magnetite microparticles caused by the reduction in the filling ratio of magnetite microparticles [27]. Thus, the Bragg reflection wavelength of 1D chain-like CC structure was found to be dependent on the microparticle concentration in suspension similar to the 3D CC structures composed of non-magnetic spherical microparticles such as silica, PS, and poly(*N*-isopropylacrylamide) hydrogel [13,27].

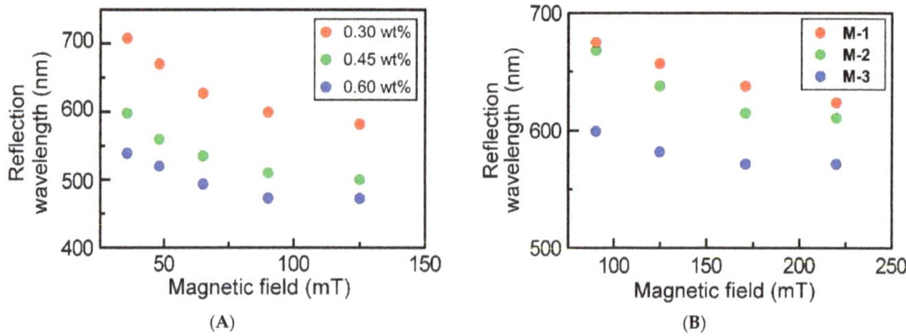

Figure 5. (**A**) Changes in the reflection peak wavelengths of aqueous suspensions of **M-3**, whose concentrations were adjusted to 0.30 wt%, 0.45 wt%, and 0.60 wt%, as a function of the external magnetic field in the range between 36 mT and 125 mT. (**B**) Changes in the reflection peak wavelengths of aqueous suspensions of **M-1**, **M-2**, and **M-3** at same concentration of 0.30 wt% as a function of the external magnetic field in the range between 91 mT to 220 mT.

As addressed in Section 3.2, one-pot hydrothermal synthesis of $FeCl_3$ adopted in this study enabled the preparation of magnetite microparticles with a diameter of 106–200 nm. Finally, we attempted to examine the dependence of Bragg reflection wavelength observed for aqueous suspension of magnetite microparticles on the particle diameter. This is because the lattice constant is dependent on the microparticle diameter. However, it is unclear in the case of 1D chain-like CC structures of the magnetite microparticles. To reveal the effect of magnetite microparticle diameters, the suspensions of **M-1**, **M-2**, and **M-3** were prepared. At this time, the concentration of each suspension was fixed to 0.30 wt%. Figure 5B shows the Bragg reflection wavelength of each suspension as a function of magnetic field strength. By comparing the Bragg reflection wavelength under the same strength of the external magnetic field in the range from 90 mT to 220 mT, the Bragg reflection wavelength shifts to shorter wavelengths as the microparticle diameter increases. For example, as applying an external magnetic field of 125 mT, the suspensions of **M-1**, **M-2**, and **M-3** showed the Bragg reflection wavelengths of 657 nm, 638 nm, and 582 nm, respectively. Such a shift of Bragg reflection wavelength can be ascribed to the difference in the M_s value. The M_s value increases with the magnetite microparticle diameter, as noted in Table 1. This suggests that larger magnetite microparticles have stronger attractive forces between microparticles, and the reflection wavelength of suspension can shift to shorter wavelengths when an external magnetic field large enough to saturate the magnetization is applied. Therefore, the Bragg reflection wavelength of the 1D chain-like CC structure of the magnetite microparticles was found to depend on the microparticle diameter.

4. Conclusions

In this study, we have established a promising methodology to synthesize monodisperse magnetite microparticles with diameters of 100–200 nm by adjusting the base concentration of the reaction solution. By applying an appropriate external magnetic field, the magnetite microparticles were immediately self-assembled into the CC structures, even in fluid aqueous suspensions, owing to their highly charged surface, thereby leading to the emergence of Bragg reflection peak in the visible wavelength range. The Bragg reflection peak could be continuously tuned from 730 nm to 570 nm by the increase in the external magnetic field from 28 mT to 220 mT. Such visible reflection characteristics of CC structures of magnetite microparticles in suspensions were found to depend on the microparticle concentration in suspension and the size of the magnetite microparticles. The present report provides the fine control of microparticle diameter as well as the investigation of these reflection properties of CC structures, which would contribute to the technological developments of full-color refractive displays and sensors by utilizing these monodisperse magnetite microparticles.

Supplementary Materials: The following supporting information can be downloaded at: https://www.mdpi.com/article/10.3390/ma15144943/s1, Figure S1: XRD patterns of the magnetite microparticles of **M-5** and **M-6** prepared by hydrothermal synthesis; Figure S2: Spatial dependence of the magnetic field strength of the NdFeB magnet used in this work; Video S1: Demonstration of reflection color changes observed for a 0.30 wt% aqueous suspension of **M-3** by applying the external magnetic field.

Author Contributions: T.S. conducted most of the experiments and wrote the original draft of the manuscript. Y.S. and N.I. analyzed the data with T.S. and N.I. revised the manuscript with T.S. and S.F. supervised this project and prepared the final version of the manuscript. All authors have read and agreed to the published version of the manuscript.

Funding: This research was supported in part by the Grant-in-Aid for Scientific Research (B) from the Japan Society for the Promotion of Science (JSPS) (Grant No. 21H02261) and the Precise Measurement Technology Promotion Foundation.

Acknowledgments: All authors deeply appreciate M. Enomoto and K. Aoki (Tokyo University of Science) for their kind support and fruitful discussion on the SQUID and zeta potential measurements, respectively, as well as T. Kato (KANETEC Co., Ltd.) for his advice on the measurements of external magnetic field using a tesla meter. A part of this work was conducted at Advanced Characterization Nanotechnology Platform of the University of Tokyo supported by "Nanotechnology Platform" of the Ministry of Education, Culture, Sports, Science and Technology (MEXT). The authors deeply thank M. Fukukawa (The University of Tokyo) for his technical support on SEM observations.

Conflicts of Interest: The authors declare no conflict of interests.

References

1. Pieranski, P. Colloidal Crystals. *Contemp. Phys.* **1983**, *24*, 25–73. [CrossRef]
2. Velev, O.D.; Kaler, E.W. Structured Porous Materials via Colloidal Crystal Templating: From Inorganic Oxides to Metals. *Adv. Mater.* **2000**, *12*, 531–534. [CrossRef]
3. Stein, A.; Wilson, B.E.; Rudisill, S.G. Design and Functionality of Colloidal-Crystal-Templated Materials—Chemical Applications of Inverse Opals. *Chem. Soc. Rev.* **2013**, *42*, 2763–2803. [CrossRef] [PubMed]
4. Fudouzi, H.; Xia, Y. Colloidal Crystals with Tunable Colors and their Use as Photonic Papers. *Langmuir* **2003**, *19*, 9653–9660. [CrossRef]
5. Furumi, S.; Fudouzi, H.; Sawada, T. Self-Organized Colloidal Crystals for Photonics and Laser Applications. *Laser Photonics Rev.* **2010**, *4*, 205–220. [CrossRef]
6. Yablonovitch, E. Inhibited Spontaneous Emission in Solid-State Physics and Electronics. *Phys. Rev. Lett.* **1987**, *58*, 2059–2062. [CrossRef]
7. Joannopoulos, J.D.; Villeneuve, P.R.; Fan, S. Photonic Crystals: Putting a New Twist on Light. *Nature* **1997**, *386*, 143–149. [CrossRef]
8. Richel, A.; Johnson, N.P.; McComb, D.W. Observation of Bragg Reflection in Photonic Crystals Synthesized from Air Spheres in a Titania Matrix. *Appl. Phys. Lett.* **2000**, *76*, 1816–1818. [CrossRef]
9. Rundquist, P.A.; Photinos, P.; Jagannathan, S.; Asher, S.A. Dynamical Bragg Diffraction from Crystalline Colloidal Arrays. *J. Chem. Phys.* **1989**, *91*, 4932–4941. [CrossRef]
10. Nakayama, D.; Takeoka, Y.; Watanabe, M.; Kataoka, K. Simple and Precise Preparation of a Porous Gel for a Colorimetric Glucose Sensor by a Templating Technique. *Angew. Chem. Int. Ed.* **2003**, *42*, 4197–4200. [CrossRef]
11. Matsubara, K.; Watanabe, M.; Takeoka, Y. A Thermally Adjustable Multicolor Photochromic Hydrogel. *Angew. Chem. Int. Ed.* **2007**, *46*, 1688–1692. [CrossRef] [PubMed]
12. Weissman, J.M.; Sunkara, H.B.; Tse, A.S.; Asher, S.A. Thermally Switchable Periodicities and Diffraction from Mesoscopically Ordered Materials. *Science* **1996**, *274*, 959–960. [CrossRef] [PubMed]
13. Iwata, N.; Koike, T.; Tokuhiro, K.; Sato, R.; Furumi, S. Colloidal Photonic Crystals of Reusable Hydrogel Microparticles for Sensor and Laser Applications. *ACS Appl. Mater. Interfaces* **2021**, *13*, 57893–57907. [CrossRef] [PubMed]
14. Hu, H.; Chen, C.; Chen, Q. Magnetically Controllable Colloidal Photonic Crystals: Unique Features and Intriguing Applications. *J. Mater. Chem. C* **2013**, *1*, 6013–6030. [CrossRef]
15. Malik, V.; Petukhov, A.V.; He, L.; Yin, Y.; Schmidt, M. Colloidal Crystallization and Structural Changes in Suspensions of Silica/Magnetite Core–Shell Nanoparticles. *Langmuir* **2012**, *28*, 14777–14783. [CrossRef] [PubMed]
16. Pal, A.; Malik, V.; He, L.; Erné, B.H.; Yin, Y.; Kegel, W.K.; Petukhov, A.V. Tuning the Colloidal Crystal Structure of Magnetic Particles by External Field. *Angew. Chem. Int. Ed.* **2015**, *54*, 1803–1807. [CrossRef]
17. Ge, J.; Hu, Y.; Yin, Y. Highly Tunable Superparamagnetic Colloidal Photonic Crystals. *Angew. Chem. Int. Ed.* **2007**, *46*, 7428–7431. [CrossRef]
18. Chi, J.; Shao, C.; Zhang, Y.; Ni, D.; Kong, T.; Zhao, Y. Magnetically Responsive Colloidal Crystals with Angle-Independent Gradient Structural Colors in Microfluidic Droplet Arrays. *Nanoscale* **2019**, *11*, 12898–12904. [CrossRef]
19. Xu, X.; Friedman, G.; Humfeld, K.D.; Majetich, S.A.; Asher, S.A. Synthesis and Utilization of Monodisperse Superparamagnetic Colloidal Particles for Magnetically Controllable Photonic Crystals. *Chem. Mater.* **2002**, *14*, 1249–1256. [CrossRef]
20. Liu, C.; Li, J. Study of Monodispersed Polystyrene Colloidal Particles Containing Fluorine. *J. Appl. Polym. Sci.* **2008**, *109*, 1604–1610. [CrossRef]
21. Seki, Y.; Shibata, Y.; Furumi, S. Synthesis of Monodispersed Silica Microparticles in a Microreactor for Well-Organized Colloidal Photonic Crystals. *J. Photopolym. Sci. Technol.* **2020**, *33*, 473–477. [CrossRef]
22. Ge, J.; Hu, Y.; Biasini, M.; Beyermann, W.P.; Yin, Y. Superparamagnetic Magnetite Colloidal Nanocrystal Clusters. *Angew. Chem. Int. Ed.* **2007**, *46*, 4342–4345. [CrossRef] [PubMed]
23. Hu, X.; Gong, J.; Zhang, L.; Yu, J.C. Continuous Size Tuning of Monodisperse ZnO Colloidal Nanocrystal Clusters by a Microwave-polyol Process and their Application for Humidity Sensing. *Adv. Mater.* **2008**, *20*, 4845–4850. [CrossRef]
24. Gao, J.; Ran, X.; Shi, C.; Cheng, H.; Cheng, T.; Su, Y. One-Step Solvothermal Synthesis of Highly Water-Soluble, Negatively Charged Superparamagnetic Fe_3O_4 Colloidal Nanocrystal Clusters. *Nanoscale* **2013**, *5*, 7026–7033. [CrossRef]
25. Yamamoto, E.; Kitahara, M.; Tsumura, T.; Kuroda, K. Preparation of Size-Controlled Monodisperse Colloidal Mesoporous Silica Nanoparticles and Fabrication of Colloidal Crystals. *Chem. Mater.* **2014**, *26*, 2927–2933. [CrossRef]

26. Lu, X.; Chen, C.; Wen, X.; Han, P.; Jiang, W.; Liang, G. Highly Charged, Magnetically Sensitive Magnetite/Polystyrene Colloids: Synthesis and Tunable Optical Properties. *J. Mater. Sci.* **2019**, *54*, 7628–7636. [CrossRef]
27. Okamoto, J.; Tsuchida, A.; Okubo, T. Colloidal Crystals Formed by Aqueous Suspensions of Monodispersed Silica, Polystyrene, and Poly(methyl methacrylate) Colloidal Spheres. *Colloid Polym. Sci.* **2011**, *289*, 1653–1660. [CrossRef]

Article

Synthesis and Spatial Order Characterization of Controlled Silica Particle Sizes Organized as Photonic Crystals Arrays

Silvia Adriana Estrada Alvarez [1,2], Isabella Guger [1], Jana Febbraro [3], Ayse Turak [3], Hong-Ru Lin [2,4], Yolanda Salinas [1,2,*] and Oliver Brüggemann [1,2,*]

1 Institute of Polymer Chemistry, Johannes Kepler University Linz, Altenberger Strasse 69, 4040 Linz, Austria
2 Linz Institute of Technology (LIT), Johannes Kepler University Linz, Altenberger Strasse 69, 4040 Linz, Austria
3 Department of Engineering Physics, McMaster University, Hamilton, ON L8S 4L7, Canada
4 Department of Chemical and Materials Engineering, Southern Taiwan University of Science and Technology, Nantai St. No.1, Tainan 71005, Taiwan
* Correspondence: yolanda.salinas@jku.at (Y.S.); oliver.brueggemann@jku.at (O.B.); Tel.: +43-732-2468-9075 (Y.S.)

Abstract: The natural occurrence of precious opals, consisting of highly organized silica particles, has prompted interest in the synthesis and formation of these structures. Previous research has shown that a highly organized photonic crystal (PhC) array is only possible when it is based on a low polydispersity index (PDI) sample of particles. In this study, a solvent-only variation method is used to synthesize different sizes of silica particles (SiPs) by following the traditional sol-gel Stöber approach. The controlled rate of the addition of the reagents promoted the homogeneity of the nucleation and growth of the spherical silica particles, which in turn yielded a low PDI. The opalescent PhC were obtained via self-assembly of these particles using a solvent evaporation method. Analysis of the spatial statistics, using Voronoi tessellations, pair correlation functions, and bond order analysis showed that the successfully formed arrays showed a high degree of quasi-hexagonal (hexatic) organization, with both global and local order. Highly organized PhC show potential for developing future materials with tunable structural reflective properties, such as solar cells, sensing materials, and coatings, among others.

Keywords: silica particles; opals; polydispersity index; photonic crystals; disLocate; Voronoi tessellations; bond order parameters

Citation: Estrada Alvarez, S.A.; Guger, I.; Febbraro, J.; Turak, A.; Lin, H.-R.; Salinas, Y.; Brüggemann, O. Synthesis and Spatial Order Characterization of Controlled Silica Particle Sizes Organized as Photonic Crystals Arrays. *Materials* **2022**, *15*, 5864. https://doi.org/10.3390/ma15175864

Academic Editors: Wei Lee, Alexander V. Baranov and Ivan V. Timofeev

Received: 13 July 2022
Accepted: 18 August 2022
Published: 25 August 2022

Publisher's Note: MDPI stays neutral with regard to jurisdictional claims in published maps and institutional affiliations.

1. Introduction

The dimensional periodic arrangements capable of controlling photo propagation are known as photonic crystals (PhC) [1]. They have been found to be responsible for the coloration involving keratin formation for a variety of animals, such as in structures found in the butterfly wings [2] or guanin crystal arrangements in chameleon skin [3]. However, perhaps the best known natural occurring photonic crystals are precious opals, which consist of highly ordered silica particles [4]. Thus, there is significant interest in silica particles (SiPs) thanks to their natural occurrence and high biocompatibility. There are reports of different methods to fabricate SiPs, however among them, the typical approach is the Stöber method [5,6]. This well-known procedure utilizes the ammonia-catalyzed hydrolysis of silicon alkoxides, specifically tetraethyl orthosilicate (TEOS), in an alcohol solution, to produce the particles.

Since the first description of the reaction, there has been considerable effort made to uncover the factors affecting the final size of the particles and the conditions for high homogeneity [7]. The temperature [8,9] of the reaction as well as the concentration of the reagents have been investigated. However, in recent years it has been discussed that a key factor to the limits of the reaction lies in the nucleation process [10], which is highly sensitive to the order and time of the addition of TEOS, pH, vortex of reaction, and viscosity of solvent, among others.

Homogeneity of the final particles can be measured in terms of the polydispersity index (PDI), where less than 0.05 is considered monodisperse [11,12]. Low PDI is particularly important for the formation of highly organized arrays such as PhC, which have become of increasing interest due to their infinite applications where tunable structural reflective properties are required, such as biological sensors, medicine [13–15], structural colored materials, textiles [16–18], electronics, and solar cells [19,20].

Research on synthetic PhCs was advanced with the work of Yablonovich and John, who engineered and detailed three types of PhCs: one, two and three dimensional [21]. After this categorization, many different approaches to produce synthetic photonic crystal were developed [22]. The primary method used is the self-assembly of monodisperse particles, such as by sedimentation, vertical deposition, physical confinement, spin coating, and most recently the dip drawing method [23]. The aim of all these methods is to yield the most hexatic lattice as possible, which in a tridimensional level translates to the face-centered cubic (FCC) or the hexagonal close packing (HCP). Both configurations have the highest packing factor of 0.74. However, the FCC is the most round-like, which improves the prospect of the photonic band gap (PBG) [24].

The aim of the present study is to use an efficient and reproducible method of silica particle synthesis, that yields low PDI, such that the particles are suitable for the formation of highly organized photonic crystals arrays. Once the PhC are obtained, we propose a quantifiable way of characterizing their degree of organization via Voronoi tessellations and spatial statistical analysis, and provide further information about the hexatic lattice.

2. Materials and Methods

2.1. Chemicals

Tetraethyl orthosilicate (TEOS, 98%), and ammonia (NH3, 28% w/w) from Alfa Aesar (Karlsruhe, Germany). Ethanol (EtOH, ABS) used for the different stages of the synthesis and work-up was purchased from ChemLab (Zedelgem, Belgium). The water (18 MΩ cm pure water) was of Milli-Q Gradient system (Millipore Corp) to keep conditions constant. For the glass slides treatment, a 1:4 piranha solution (H_2O_2/H_2SO_4) was used. Hydrogen peroxide (33.3%) was purchased from VWR Chemicals and sulphuric acid (95–97%) was received from Merck. All reagents were used as received.

2.2. Instruments for Syntehsis and Sample Preparation

During the fabrication of the silica particles (SiPs), a syringe pump (KDS 100 Legacy Syringe Pump, KD Scientific Inc., sourced from Merck KGaA, Darmstadt, Deutschland) was used for the controlled addition rate of TEOS. After 3 h of stirring, the solution was centrifuged (Thermo Heraeus Multifuge 1S Centrifuge, Thermo Fisher Scientific Inc., Linz, Austria) for 15 min at 5000 rpm, and the same settings were used for washing the particles. An ultrasonic cleaner (with digital timer—VWR International, Vienna, Austria) was used during the different steps of the synthesis and the characterization, to fully disperse the particles. Before analysing the samples under the scanning electron microscopy (SEM), the fabricated silica particles and photonic crystals were gold coated (Sputter Coater and Carbon Coater for TEM/SEM, S series, Instrument Futurism, Linz, Austria) with an average layer of 7 nm.

2.3. Characterization Methods

Dynamic light scattering (DLS) (Zetasizer Nano ZSP from Malvern Instruments, Worcestershire, UK) measurements were carried out to determine the average hydrodynamic diameter (D_h) and the polydispersity index (PDI). In this study, the PDI refers to the uniformity of the particles (where a value close to 0.0 is a perfect homogeneity of the sample, while a 1.0 is a sample with a large variety of sizes). The samples were prepared in 0.1 M solution with Milli-Q water as dispersant in disposable cuvettes (DTS 0012), sonicated for 30 min. The DLS measurements via intensity were performed at 25 °C without filtration before the measurements. SEM images were obtained with a Jeol 6400 (Jeol, Peabody, MA, USA). The optical observation of the SiPs was carried out with an optical microscope at ×200 magnification (ZEISS axio imager A1m—Carl Zeiss AG, Vienna, Austria). The reflection spectra were measured using a spectrometer (LAMBDA 1050+ UV/VIS/NIR Spectrometer with 150 mm InGaAs Int. Sphere and UV WinLab software—PerkinElmer Inc., Linz, Austria). To characterize the organized silica arrays, the same coating conditions and SEM were used. The degree of organization was determined with Voronoi tessellation, using Mathematica package **disLocate**, a suite of tools to rapidly quantify the spatial structure of a two-dimensional dispersion of objects [25].

2.4. Synthesis of SiPs

The SiPs were synthesized using the Stöber method [5], through the controlled hydrolysis and condensation of tetraethyl orthosilicate in an alcohol medium. With the purpose of maintaining all conditions uniform for all experiments, the only variant to the synthesis of the different sizes of SiPs was the amount of solvent used, i.e., ethanol, while the molar amount of ammonia and of water was kept constant in relation to the used TEOS. The amount of EtOH used were 100 mL, 80 mL, 60 mL, and 40 mL.

The different amounts of ethanol were mixed with 8 mL of ammonia (80 mmol) and 3 mL of water in a 250 mL round bottom flask at 60 °C for 10 min. 6 mL of TEOS (26 mmol) were added slowly with the addition pump set to 18 mL/h. Constant and controlled addition speed allows reproducibility of the particles size. After the full addition of all reagents, the reaction was stirred for 3 h and then stopped by centrifugation to prevent further nucleation. It was then washed three times with 50 mL of EtOH to remove all unreacted monomers. This synthesis yielded the-so-called samples SiPs100, SiPs80, SiPs60, and SiPs40, in relation to the amount of solvent used during their preparation.

2.5. Preparation of SiPs Based PhC

The opalescent PhC were prepared by means of the evaporative deposition self-assembly method using the suspension of the SiPs in EtOH. To start with, the glass slides were made hydrophilic by immersing them into a 1:4 piranha solution (H_2O_2/H_2SO_4) while stirring the solution for 1 h. The glass slides were then rinsed with several aliquots of Milli-Q water to remove the piranha solution. The slides were then dried with a stream of nitrogen. This treatment further activates the -OH groups on the surface of the glass.

The glass slides were then inserted vertically in conical-bottom centrifuge tubes. Figure 1 exemplifies the formation of the opalescent films on the piranha solution treated glass. These tubes contain colloidal SiPs suspensions of the different concentrations (0.5%, 1%, 2% and 4% *w/w* of the SiPs). The conical shape of the tubes was necessary to prevent a thicker formation of opal on the bottom of the slide. This is expected to occur since the concentration of SiPs increases as the solvent evaporates. After the ethanol was evaporated completely, the SiPs organized array was obtained on both sides of the glass slides, which appeared as an opalescent structure.

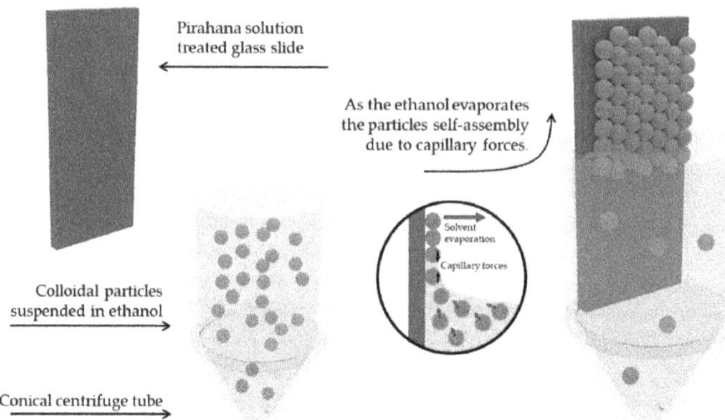

Figure 1. Setting to obtain organized an opalescent array of SiPs. The particles were fully suspended in ethanol at different concentrations, within a conical bottom centrifuge tube. The conical bottom was necessary to avoid a drastic thickening of the arrays at the bottom of the slide. A piranha solution treated glass slide is carefully placed vertically in the tube. The system is placed in a quiet, vibration free environment, where the ethanol evaporates at constant rate. The suspended particles self-assemble on the glass due to capillary forces in an organized manner as the evaporation takes places.

3. Results and Discussions

To begin with, we synthesized a library of different sizes of silica particles. To keep control of the reproducibility, only the solvent amount was changed to obtain different silica particle sizes [26]. The SiPs were characterized via SEM and DLS, where the diameter and the hydrodynamic diameter (D_h) were obtained; the latter here is usually a larger number, since it refers to moving particles in a solution. To ensure the quality of the subsequent PhC, only samples with PDI lower than 0.05 were considered. For the formation of the opalescent organized arrays, the particles of different sizes were dispersed in ethanol, which was later evaporated leaving the arrays of sedimented particles. This material was characterized by scanning electron microscopy (SEM). Additional to this characterization method, we analyzed the lattice using Voronoi diagrams, with the purpose of obtaining a quantifiable method of degree of organization.

3.1. Synthesis and Characterization of SiPs

By maintaining all conditions uniformly in all experiments and only varying the amount of solvent used, the reproducibility capability was maximized. This includes the use of the same oval stirrer as the different vortexes affected the nucleation and thus the final average particle size. The average SiPs diameter size was measured via SEM, and the average hydrodynamic diameter was measured via DLS. For significant validation of the results obtained from the SEM, 50 different points were measured of 5 different samples of SiPs for each experiment. In the case of DLS, 5 aliquots were measured from 15 different samples.

As an example of the sensitivity of the reaction, Figure A1, of the appendix shows the hydrodynamic diameter obtained using 100 mL of solvent and a 20 mm oval stirrer as on average 135.6 ± 2.5 nm, however using a 25 mm oval stirrer the hydrodynamic diameter is 178.8 ± 0.6 nm. All the experiments carried out in this research were performed using a 25 mm oval stirrer.

The varying amounts of ethanol used were 100 mL, 80 mL, 60 mL, and 40 mL, which yielded an average particle diameter of 150 ± 3.1 nm, 233 ± 2.6 nm, 365 ± 2.4 nm, and 490 ± 4.3 nm, and a hydrodynamic average diameter (D_h $\overline{X}$) of 178.8 ± 0.6 nm, 242.7 ± 2.2 nm, 396.3 ± 2.3 nm, and 518.5 ± 4.2 nm for SiPs100, SiPs80, SiPs60, and SiPs40, respectively. Table 1 summarizes the experiments carried out, together with the results obtained from measuring the diameter (ø) and hydrodynamic diameter (D_h).

Table 1. Impact of the different amounts of EtOH used for the synthesis of SiPs on the diameter.

Experiment Code	EtOH (mL)	Average Diameter [1] ø (nm)	Hydrodynamic Average Diameter [2] D_h (nm)	PDI [2]
SiPs100	100	150 ± 3.1	178.8 ± 0.6	0.002
SiPs80	80	233 ± 2.6	242.7 ± 2.2	0.020
SiPs60	60	365 ± 2.4	396.3 ± 2.3	0.022
SiPs40	40	490 ± 4.3	518.5 ± 4.2	0.034

[1] Measured via SEM. [2] Measured via DLS.

The scheme in Figure 2a shows the reaction that takes place when synthesizing the SiPs from TEOS. After the start of the addition of TEOS to the reaction, hydrolysis takes place to form silanol monomers, which later condense giving way to a siloxane network, which allows the formation of nuclei and subsequently growth. This step takes place during the 3 h that the reaction was stirred. By consistently stopping the reaction at a set time, further formation of nuclei and growth is prevented, thus controlling the size of the synthesized SiPs while keeping a low PDI. Figure 2b represents the inversely proportional relationship that the amount of solvent has to the final size of particles and shows the respective SEM images.

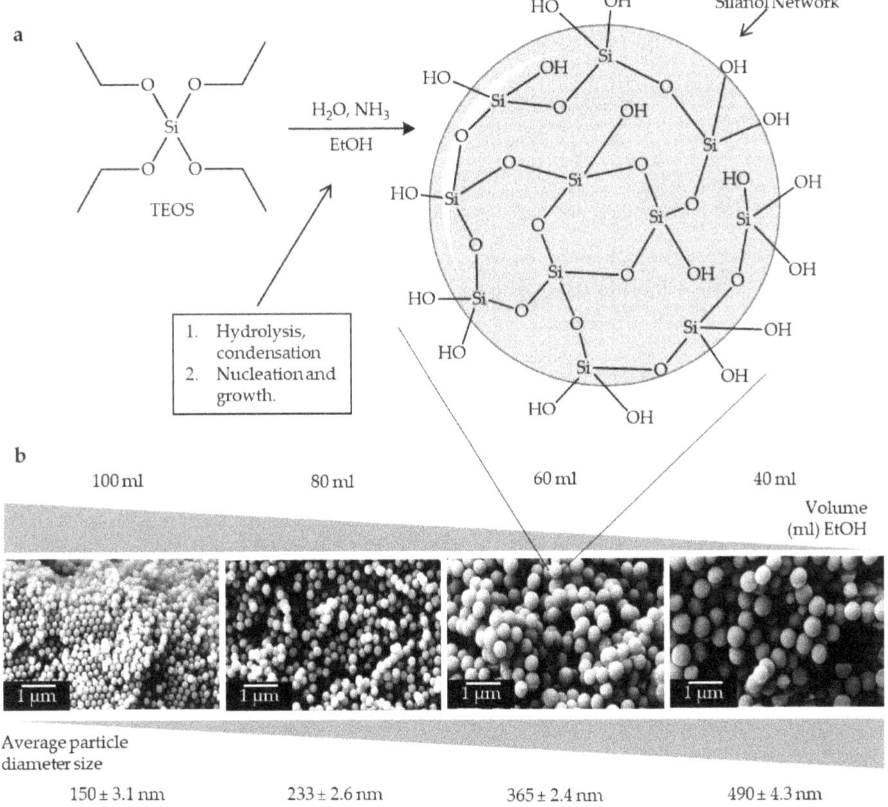

Figure 2. *Cont.*

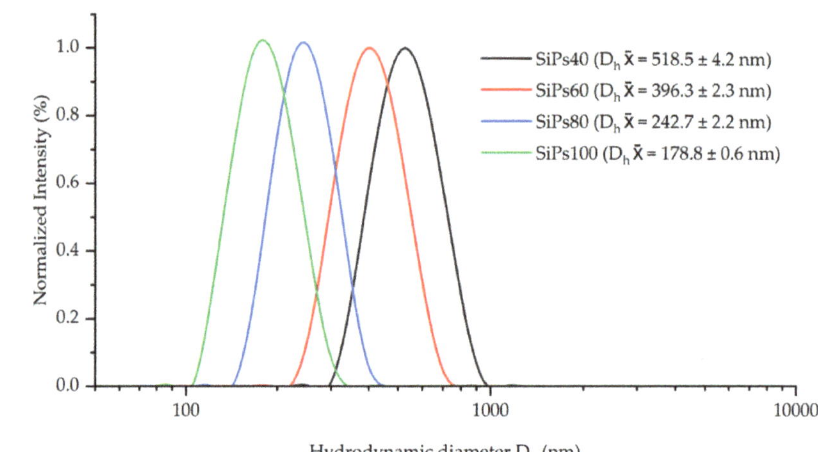

Figure 2. Synthesis and characterization of the SiPs. (**a**) Scheme of the SiPs synthesis. After the adhesion of TEOS to the ethanol solution, hydrolysis takes places promoting the formation of the silanol network. This in turn starts the nucleation for the formation of the SiPs. (**b**) SEM images of the obtained SiPs, and the representation of the inversely proportional relationship between the amount of EtOH and the final diameter of the SiPs. (**c**) Hydrodynamic sizes of the SiPs measured via DLS. The hydrodynamic diameter is usually a higher value than the diameter since it refers to moving particles in a solution.

A monodisperse sample of SiPs is necessary for a highly organized lattice; this required the PDI to be lower than 0.05. This index was measured together with the D_h using DLS. The results show that the previously explained procedure for the synthesis of the SiPs yields particles with the desired range of PDI. All samples used for the consecutive organizing sedimentation showed a PDI between 0.002 and 0.05. Figure 2c shows the graphs obtained from the DLS analysis displaying the D_h and the PDI.

3.2. Formation and Characterization of the SiPs PhC

The opalescent organized arrays were formed using different initial concentrations of SiPs in ethanol (0.5%, 1%, 2%, and 4% *w/w*). The solvent evaporation method used for the formation of the opals on the glass slides (2 × 2 cm), translated to a constant increase of the suspended particles concentration as the solvent evaporated, because of -this, all the samples were thicker on the lower part than on the upper part. Therefore, the thickness considered for this study was only taken from the middle of the sample.

The following two subsections explore the characterization of the PhC via SEM and the degree of organization with the help of statistical spatial order parameters.

3.2.1. Visual Characterization

SEM was used for characterizing the top and side view of the formed PhC. The top view shows the morphology and the high degree of organization. Visually it is possible to recognize the hexagonal pattern formed by the SiPs (shown in Figure 3 on the SEM top view of the SiPs40). This is a characteristic trait of the FCC and HCP arrangement. This is also supported by the coordination cloud or fast Fourier transform (FFT) of the image, pictured in the insert in the top right-hand corner (Figure 3a), showing six distinct equally spaced lobes as expected for good hexagonal close packing. SiPs60 have less distinct lobes, suggesting that it is less well ordered than the other three films.

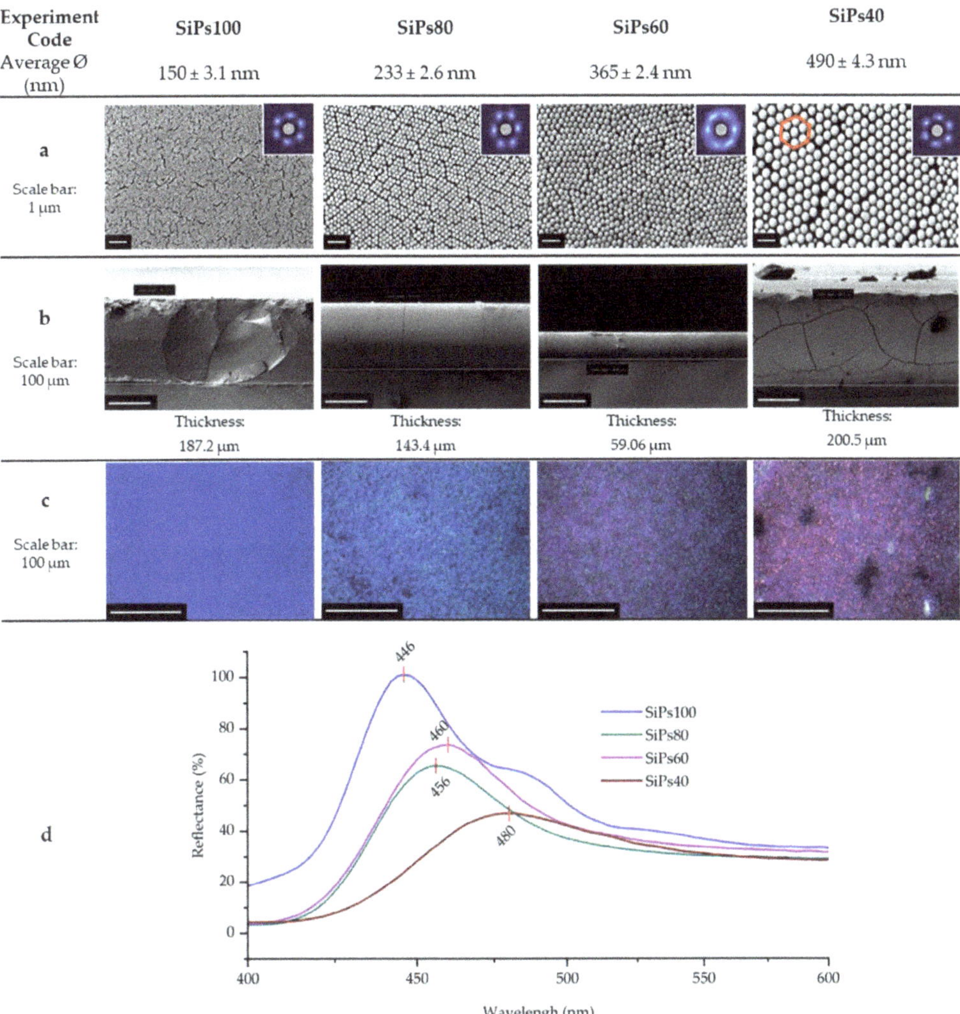

Figure 3. SEM and optical microscope observation of the SiP photonic crystals and their reflected wavelengths. The samples used for these figures were made using 1% of suspended particles in EtOH via solvent evaporation method. (**a**) SEM top view of the silica particle based photonic crystals, the inserts in the images show the correlation map or fast Fourier transform. These self-assembled organized layers were formed after the evaporation of the solvent. Qualitatively, they show a high level of hexagonal organization, as marked in the red frame. (**b**) SEM side view of the formed PhC. The thickness of the layers does not seem to follow any trend related to the particle size. (**c**) Optical microscope view of the same samples showing different colors. (**d**) The reflected wavelength that aligns with the results of part c. The wavelength of SiPs80 and SiPs60 are very close, which explains the similar coloration seen in the optical microscope.

These kinds of organization are the most desirable for photonic crystals since a close packing factor is most suited for wavelength propagation. A further examination and quantifiable reasoning of the hexagonal configuration of the SiPs is found on the next section.

The thickness of the formed opals was not found to follow a specific pattern and did not seem to be influenced by the particle size. Our reasoning to this occurrence is that after a certain number of stacked layers, the integrity is compromised due to weight causing the most recently sedimented layers to slide down possibly bringing some of the older sedimented layers with them. This occurrence also explains the source of cracks seen specially on the bigger particles (Figure 3b, SEM side view of the SiPs40).

By placing the samples in an optical microscope, it was possible to observe the reflected wavelength, of which the source is the microscope beam. The smallest particles (SiPs100) reflected an intense homogeneous blue, similar to what is commonly known as "cobalt blue". The second smallest particles showed shades of green combined with domains of blue, giving up a color that could be describe as "dark teal". The SiPs60 give up a more purple like appearance as it shows domains of blue, and at the same time of red, this could be recognized as a "cyber grape" appearance. The biggest particles reflected a darker red like tone that could be seen as a "plum" color. Similar visual coloration was obtained by W. Gao [26], however their method of self-assembly was gravitational sedimentation, instead of evaporation deposition.

This color observation matched the results obtained from the UV/Vis spectrometer, shown Figure 3d. For this characterization, the samples were place horizontally in the machine for the measurements. The closeness of the measured wavelengths of SiPs60 and SiPs80 is attributed to the blue domains found in both samples.

3.2.2. Spatial Statistics Characterization

To describe quantitatively the structure and assign a value to the relative order, the formed PhC arrays were characterized from SEM micrographs with spatial statistics obtained using the Mathematica package, **disLocate** (**D**etecting **I**ntermolecular **S**tructure **L**ocate**d** at particle positions) [25].

Order on the local scale can be sub-categorized into three distinct types: translational, entropic and angular. Translational order occurs when every particle in the system has an exact position that repeats at a specific distance, defined by a specific translation period.

Entropic order is achieved when the amount of free volume (sections unoccupied by particle mass) encompassed by the system is the lowest possible such that the system is at maximum density. Angular order is related to the relative arc-separation of the "bonds" or touching contacts between a particle and its neighbors.

Complete periodicity is seen when the neighbors for any particular particle have the same angular arc symmetry, as well as being equidistant, with maximized covering area due to the equivalent position of each particle relative to all others. If any one of these types of order are not met, the system can be considered to be in a mesophase [27], such as that observed for plastic crystals (limited orientational or rotational order, but long-range translational order [28]) or liquid crystals (limited translational order but long-range angular order) [29].

To assess each type of order for the silica particles opal arrays, different spatial order metrics were applied to the images, as shown schematically in Figure 4.

Local free volume and the complementary metric, covering area [30] can be calculated by partitioning the substrate into a Voronoi tessellation around each individual particle (Figure 4a,d) [31]. To define each cell of the tessellation, a perpendicular line at the midpoint along the line-of-sight vector connecting nearest neighbors around every particle is calculated. The intersection of these lines defines the unit cell surrounding each particle. The covering area is then extracted and compared to that expected for a hexagonally close-packed system, to show the distribution of hexagonally packed regions. As there is an entropic driving force that aligns faceted or functionalized particles so as to maximize the system entropy by minimizing the free volume, analogous to chemical valence states [32–34], it is possible to define a coordination number from the number of Voronoi cell facets which contain the particle (see Figure 4d). To show the various local order states, the Voronoi tessellation maps can be colored by the cell area deviation from that expected for perfect hexagonal packing, by the coordination number, or by the deviation of the bond order parameter from the expected symmetry value (discussed below).

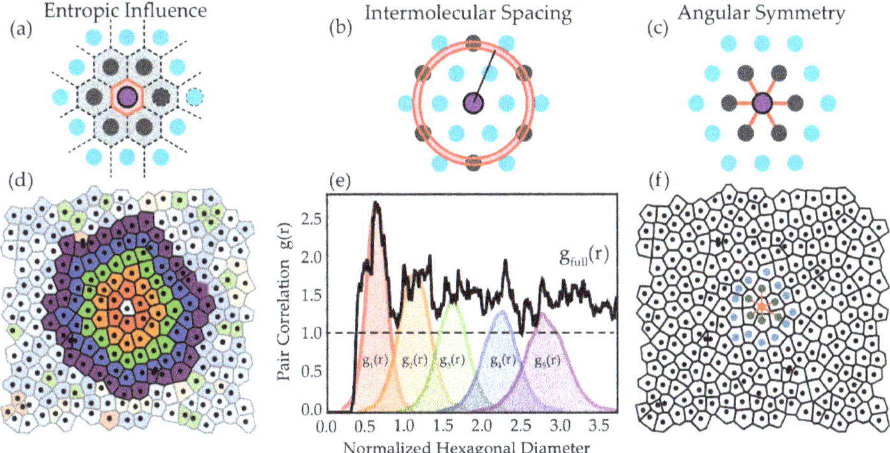

Figure 4. Schematic of order metrics used to characterize the morphology located at particle positions: (**a**) Voronoi tessellations (**b**) pair correlation function (**c**) bond order parameter, for each category of order possible in a 2D pattern. (**d**) Voronoi tessellation of a planar pattern of particles with coloring scheme representing the Voronoi shell description of neighbors. The coloring scheme of outlier cells reflects the deviation of covering area from that of a hexagonal lattice. (**e**) The pair correlation function separated by the Voronoi shells, showing the connection between local and global order metrics. The sum of finite shellular descriptions leads to an approximation of the configurational average g(r). At large values, the value of the pair correlation goes to uniform probability (a value of 1 shown by the dotted horizontal line). (**f**) Voronoi tessellation of a planar pattern of particles showing the "bonds" connecting adjacent particles, where the angle between bonds compared against the expected angle for hexagonal packing can be used to determine the local bond order for each cell, as well as the ensemble average global bond order.

Going beyond the first neighbor, it is possible to quantify the positional order probabilistically using the pair correlation function [35]. The pair correlation function *g(r)* describes the number of objects within a small shell at a distance away from a central particle and averaged over all particles in the system (Figure 4b,e). Imagining the neighbor distances as a series of Voronoi cells radiating out from each particle, the ensemble sum of each Voronoi cell leads to an approximation of the configurational average radius *(r)*. In a periodic lattice, the objects are evenly spaced out at specific distances by lateral translations of a repeated unit (unit cell). The degree of order can be determined by the number of peaks in the pair correlation function, with predictable spacing between peaks for each Bravais lattice. At large distances, the pair correlation function converges to uniform probability (shown by the horizontal dotted line in Figure 4e). The convergence of the pair correlation function is related to the extent of spatial periodicity, with high periodicity showing many multiple shells before there is equal chance of finding a particle at any distance. By contrast, a system with complete spatial randomness would have a uniform probability for all values of distance, above the nearest neighbor distance (or particle diameter for particles in contact).

Angular orientation order describes the likelihood of finding an object at a given angular arc-separation between neighboring particles, most commonly thought of as the symmetry state of a system. Angular order can be calculated by using the bond order parameter [36], which compares the angle between the central particle and its closest neighbors against a specified set of symmetry basis vectors (Figure 4c,f). In the aggregate across all particles, this can be collapsed down into a coordination cloud or fast Fourier transform (FFT) of the image [37,38]. For close-packed systems of circles, hexagonal configurations have the highest density, so the 6-fold bond order values (q_6) are the most crucial ones. All of these methods have been widely used to characterize disorder, identify polycrystalline and disordered sections, and extract the probability of intermolecular spacings [39,40]. The complete spatial statistic description for each of the formed PhC arrays are summarized in Figures A2–A5.

Figure 5 shows the pair correlation function for the particle centroids for the SiPs arrays. As can be seen in Figure 5a, the pair correlation functions are misaligned for the various arrays, confirming that the average spacing between particles is inversely correlated with the amount of solvent addition. Normalizing the distance to the average spacing ($2r_{hex}$) of a hexagonal lattice with the same number density of particles (located at particle positions), as seen in Figure 5b, collapses the distributions into a shared spatial reference frame where peak positions can easily be compared.

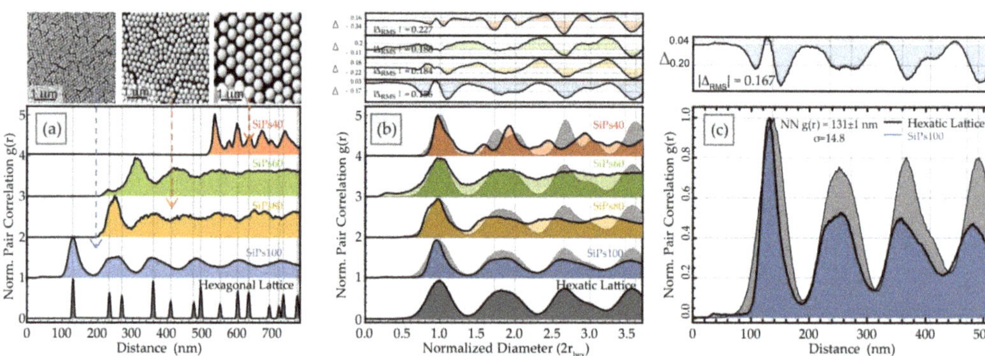

Figure 5. Pair correlation functions *g(r)* of centroids obtained from SEM images of PS inverse opals. Colored arrows show the corresponding SEM image resulting in the pair correlation function (**a**) Measurements of objects shows that the peak positions are misaligned. This confirms how the average spacing between particles is inversely correlated with the amount of solvent addition. The expected *g(r)* for a hexagonal lattice with similar intensity (average particles per unit area) as SiPs100 is shown for comparison. (**b**) Normalizing the distance to the average spacing ($2r_{hex}$) of a hexagonal lattice with the same number density (located at particle positions) collapses the distributions into a shared spatial reference frame where peak positions can easily be compared. The smallest and largest diameter particles show correlation with the hexagonal lattice of similar intensity (average particles per unit area) for more than 4 degrees of order (peaks in the correlation function). For the intermediate diameters only one or two degrees of order are observed, indicating a larger amount of disorder in those systems. (**c**) The differences in *g(r)* are shown as the grey sections on an overlay of pcf from SiPs100 particle distributions and that of the matched hexatic lattice. The widths of these peaks correspond to the average mean displacement each particle has relative to this expected spacing. The inserts above (**b,c**) show the difference spectrum, which is the subtraction between a pair correlation function and the hexatic lattice.

For SiPs80 and SiPs40, the nearest neighbor distance (maximum of the first peak of the pair correlation) is slightly larger than the average particle diameter (252 ± 2.3 nm and 531 ± 1.6 nm, respectively), consistent with the visible cracks observed in various regions increasing the average distance between particles.

For SiPs60, where there are no visible cracks, this spacing is smaller than the particle diameter (327 ± 3.5 nm), and there is even a small feature before the first peak around 100 nm. Typically for touching close-packed particles in a single layer, it is not possible for the particles to be closer than the particle diameter, and the pair correlation function is zero for values of distance less than the mean particle size. These features for SiPs60 suggest that there is some vertical disorder in the film, and what is visible in the SEM image is more than the top layer of the film. This is supported by the significantly thinner film thickness observed for this solvent addition. With densification of the film, particles have greater disorder vertically, and appear to pack more tightly. Overall, this film appears to be more disordered, with the pair correlation function having only two broad peaks before becoming featureless and converging on uniform probability. This type of behavior is typically associated with mostly disordered films, with a small degree of short-range order. This is also consistent with the coordination cloud or fast Fourier transform (FFT) shown in Figure 3, where the six lobes showing hexagonal packing are smeared, suggesting a loss of angular order. The higher level of disorder in this film could explain the inhomogeneous coloration for SiPs60, with both red and blue patches and even darkened regions visible. The periodicity is disrupted, and the reflected color is non-uniform.

Though SiPs100 also have an average spacing (131 ± 1.1 nm) less than the predicted particle diameters, the features of the pair correlation function are very different. Unlike for SiPs60, this film shows good correlation with a hexagonal lattice of similar intensity (average particles per unit area) for more than 6 degrees of order (peaks in the correlation function). The average spacing is consistent with that of a hexagonal lattice, as shown in Figure 5a. However, there is some positional disorder from a perfect hexagonal close packing, shown by a broadening of the peak widths. Slight deviations of the particle positions close to the global lattice positions will not change the mean particle spacings, but the small fluctuations increase the chance for neighbors to be found at slightly different distances, yielding peak broadening. Using our tool set, we can determine the extent of these fluctuations using an "hexatic" lattice, determined through an ensemble average of a hexagonal lattice matched to the particle spacing which has been modified by randomly displacing the particles from their center with a mean distance relative to a normal probability distribution [23]. In this way, the lattice retains the angular symmetry of the hexagonal lattice but relaxes the condition of maximal density. An overlay of the pair correlation function from SiP100 particle distributions and the matched hexatic lattice is shown in Figure 5c. The widths of the peaks correspond to the average mean displacement each particle has relative to its expected spacing. The insert above shows the difference spectrum, which is the subtraction between a pair correlation function and the hexatic lattice. These comparisons are also shown in Figure 5b. For films other than SiPs60, the pair correlation function shows good correlation with a hexatic lattice, with SiPs40 also showing a high degree of order with many peaks in the correlation function.

A critical feature of the hexatic lattice is that each particle has six-fold symmetry, with six nearest neighbors. This local order can be observed with Voronoi tessellations of the films (see Appendix A Figures A2–A5). All four systems show greater than 70% regions with Voronoi cells of high hexagonal entropic order (uncolored cells in Figure 6), within a larger area that is not globally perfectly hexagonal, (as already seen from the pair correlation analysis). This suggests roughly hexagonal symmetry arrangements, but with some relaxation of strict hexagonal packing. With defects or local disorder, the steric frustration at the boundaries between large hexagonal sections typically forms pairs of Voronoi cells with alternating 5 and 7 sides (green and orange regions). These are the so-called "disclination" defects [7], where the local number of neighbors is violated. In an analogy with dislocations, which are positional defects, a collection of 5/7 declination pairs are used to define "grain boundaries" between entropically ordered sections of 2D colloidal crystals [41,42]. The visible spacing or cracks between particles in certain regions for the films do not correspond well to the 5/7 declination pairs. This suggests that the cracks are not low-angle grain boundaries. They disrupt the periodicity of the lateral spacing (i.e., are not close-packed) but retain high hexagonal order, as would be expected for a hexatic system.

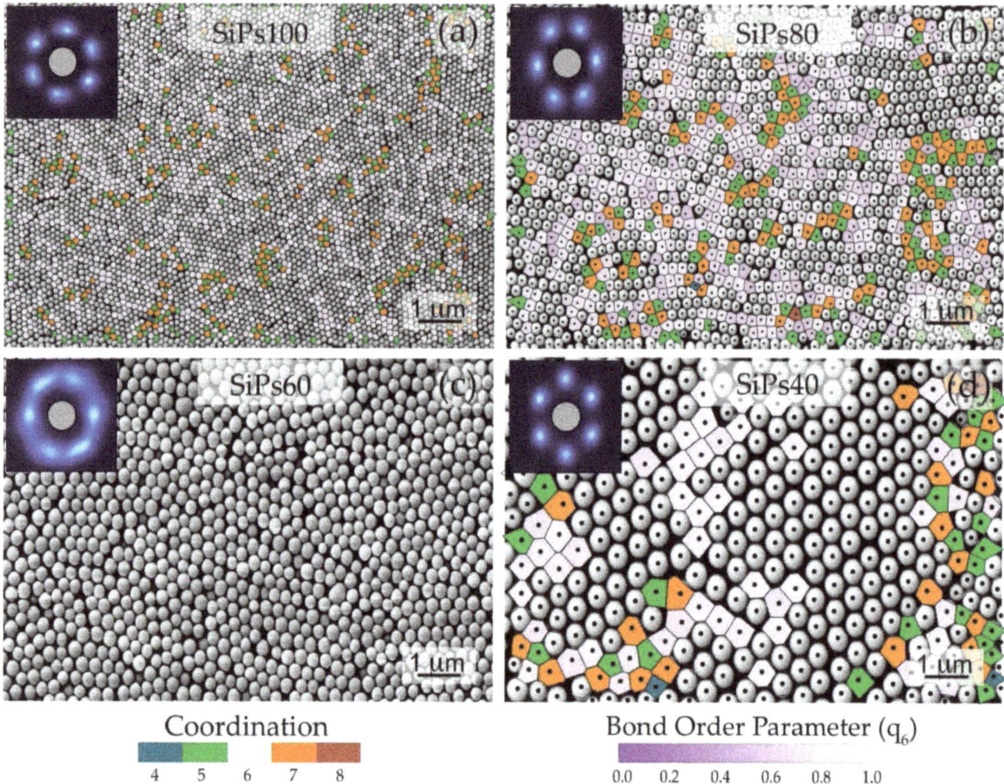

Coordination

4 5 6 7 8

Bond Order Parameter (q_6)

0.0 0.2 0.4 0.6 0.8 1.0

Figure 6. Overlay of Voronoi tessellation of the particle centroids over the corresponding SEM image of the particles for (**a**) SiPs100 (**b**) SiPs80 (**c**) SiPs60 and (**d**) SiPs40. Cells are colored by the number of shared facets and by the normalized bond order parameter for the type of symmetry for six neighbored cells. Whiter areas indicate particles that have high angular order in that symmetry basis. Particles with 5 and 7 neighbors typically highlight the grain boundaries and dislocation lines in their respective symmetry basis, but are not consistent with the cracks visible in the SEM images. Incorporating normalized bond order between 60% and 90% of that expected for a hexagonal lattice for 6 neighbor domains highlights the cracking observed in the films. SiPs60 in part c showed no correlations.

The bond order showcases the regions that have the correct angular spacing but lack translational symmetry, i.e., that are hexatic rather than perfectly hexagonal. The global q_6 bond order yields the average hexagonal nature of the film as a whole, and the local bond order shows how close each Voronoi cell is to being perfectly hexagonal. For the SiPs films, the global bond order parameter scales as SiPs60 < SiPs80 < SiPs100 < SiPs40. The values for SiPs100 and SiPs40 are similar, likely to within error. This behavior is supported by the pair correlations shown in Figure 5b, where both SiPs100 and SiPs40 showed regular peaks, with good correlation to a matched hexatic lattice, though SiPs40 seemed less so. This can be explained through the Voronoi analysis, where there is a region on the right-hand side of the image which contains a higher number of defects than the rest of the film. As the particle density is much lower for the SiPs40 film compared to SiPs100, the mean global values are more affected by such outliers with a smaller overall number of particles. For systems of similar intensity, the global values would be expected to be similar for both films. The lower value of global bond order for SiPs80 is likely due to the presence of more cracks in the system (as discussed below). The outlier is the SiPs60 film, which is the least hexatic (only 73% as much as a pure hexagonal system).

The localized bond order of each particle is lower in the crack regions, separating highly ordered domains where the local bond order is higher (white domains in Figures A2, A3, A4 and A5e,f). For SiPs100 and SiPs40, the local bond order between 0.8 and 0.9 roughly follows the regions where cracks are visible in the film, when combined with the 5/7 neighbor domains, as shown in Figure 6. These sections separate the larger domains where there is almost perfect hexagonal packing. SiPs40 appears slightly more ordered than SiPs100 (crack domains 33% vs. 39%, respectively), but this value could be skewed by the low number of particles. Likely within error, both structures are similarly ordered.

The SiPs80 film requires a combination of 5/7 domains plus 6-fold domains with local bond order down to 60% of that expected for a pure hexagonal lattice to accommodate (imperfectly) the cracks. These together with the 5/7 declination pairs make up 56% of all domains. There seem to be more cracks visible in this film compared to SiPs100 and SiP40s, and those crack regions are more disordered.

The SiPs60 film has no correlation, as the overall film is disordered, rather than ordered with cracks. This is also supported by the FFT (coordination cloud) which is smeared out for SiPs60, compared to the bright six-fold distinct lobes for the other solvent additions.

Overall, the spatial statistical analysis of the different opalescent films suggests that SiPs100 and SiPs40, with the smallest and largest diameter particles, are similarly ordered. They consist of large hexatic domains, separated by cracks in the film that are not acting as grain boundaries, but show some loss of angular symmetry. The SiPs40 film is slightly less close-packed, with the average spacing between particles larger than the measured particle diameters. SiPs80 is less ordered, with more visible cracks and a greater number of Voronoi domains that have both relaxed translational and angular periodicity. However, the film is uniform with hexatic ordered domains between the cracked regions. SiPs60 has some short-range order but lacks both translational and angular periodicity over much of the film. Unlike the other SiPs films, it is likely also disordered vertically, without clear layered lattice-like planes, as the SEM image from the top of the film likely shows more than one layer, resulting in a thinner film and inhomogeneous opalescence.

4. Conclusions

The different sizes of silica particles were synthesized by only changing the amount of solvent used. This suggests that, although the reaction between ammonia and TEOS remains the same, the concentration of both reagents is changed. This, in turn, caused the growth process to be affected. In the case of 100 mL of EtOH used, there was a large nucleation, but the growth was diminished due to the distances between nuclei. In contrast, with a higher concentration, as in the case of the 40 mL of EtOH, the growth was enhanced due to the proximity of the forming silanol network. By closely monitoring the rate of addition of TEOS, we promoted a constant supply of the reagent and a controlled consumption of the same. This was the substantial reason for the low PDI, which is a key point for the formation of the PhC.

During the formation of the organized arrays on the piranha-solution-treated glass slides, it was of key importance that the evaporations took place in quiet, vibration free environment. These little disturbances cause wavelike shapes to appear on the dried PhC, which showed different thickness and proved unreliable for characterization. The assembly of the particles on the slides is largely due to the capillary forces at the meniscus area of the evaporating solvent and the glass slide.

The spatial statistics of the organized arrays yielded a quantifiable comparison of the degree of spatial, angular, and entropic organization. By simple visual observation, it is possible to perceive some hexagonal packing, typical of FCC and HCP lattices, however it is very difficult to determine to what extent and be able to compare different pathways of organizing particles, especially when the experiments are similar, and the positional differences are similar. In this study, we were able to observe that both 100 mL of EtOH and 40 mL of EtOH addition yielded highly organized films, with good correspondence to a quasi-hexagonal (hexatic) lattice, where some positional symmetry is relaxed, but with high angular symmetry, resulting in good opalescent films. The addition of 60 mL of EtOH resulted in a more disorganized film, with only 73% the character of a hexagonal close-packed lattice, with some loss of periodicity in the vertical direction as well, yielding non-uniform reflectance. The use of such statistics can drive the formation of highly ordered self-assembled arrays.

Author Contributions: Conceptualization, S.A.E.A.; methodology, S.A.E.A.; software, A.T.; validation, S.A.E.A., Y.S., H.-R.L. and O.B.; formal analysis, S.A.E.A., Y.S., J.F. and A.T.; investigation, S.A.E.A. and I.G.; resources, O.B.; data curation, S.A.E.A. and I.G.; writing—original draft preparation, S.A.E.A.; writing—review and editing, S.A.E.A., Y.S. and A.T.; visualization, S.A.E.A. and A.T.; supervision, Y.S., A.T, H.-R.L. and O.B.; project administration, O.B.; funding acquisition, O.B., H.-R.L. and S.A.E.A. All authors have read and agreed to the published version of the manuscript.

Funding: This research was funded by Linz Institute of Technology (LIT), Johannes Kepler University Linz, project number LIT413760001 CHAMS, and Natural Science and Engineering Council of Canada, project number RGPIN—2019-05994, and the Ontario Ministry of Research and Innovation project number ER15-11-123. The open access was provided by Johannes Kepler Open Access Publishing Fund.

Institutional Review Board Statement: Not applicable.

Informed Consent Statement: Not applicable.

Acknowledgments: P. Oberhumer and K. Stadlmann from the Centre for Nano and Surface Analytics (Zona, JKU) for their support in SEM analysis and gold coating. C. Fiedler for her support in scheme design. The open access was provided by Johannes Kepler Open Access Publishing Fund.

Conflicts of Interest: The authors declare no conflict of interest. The funders had no role in the design of the study; in the collection, analyses, or interpretation of data; in the writing of the manuscript, or in the decision to publish the results.

Appendix A

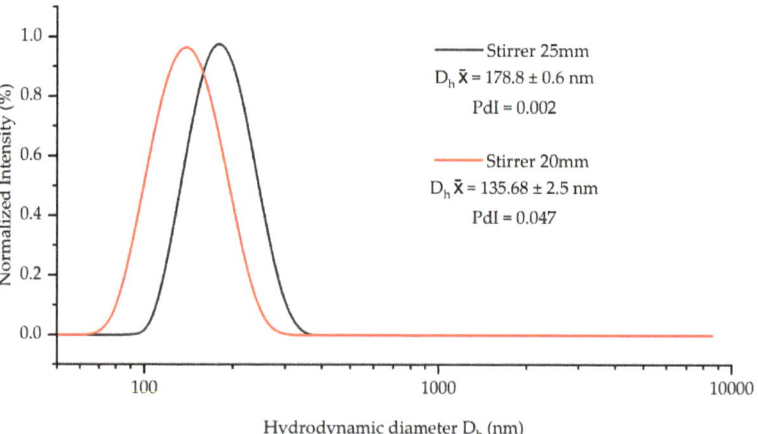

Figure A1. Hydrodynamic average diameter of particles measured via dynamic light scattering (DLS) using two different stirrer sizes. The different samples were fabricated with two sizes of stirrers, 20 and 25 mm. The vortex created by the stirrers resulted consistently in different particles sizes, although all the other conditions were carefully maintained constant.

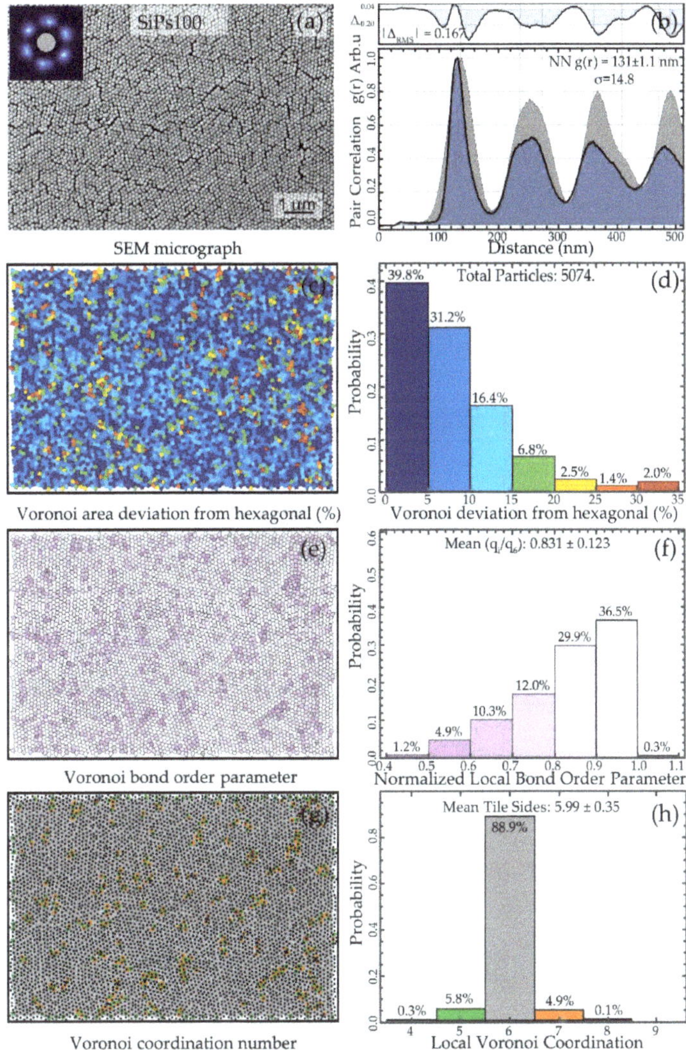

Figure A2. Spatial statistics of sedimented SiPs100. (**a**) SEM image with insert of coordination map or fast Fourier transform (FFT) showing six distinct lobes as expected for hexagonally packed systems. (**b**) pair correlation function for the particle distributions (blue) and that of the matched hexatic lattice (grey). Using the average spacing ($2r_{hex}$) of a hexagonal lattice with the same number density (located at particle positions), the matched hexatic lattice is determined by randomly displacing the particles from their center with a mean distance relative to a normal probability distribution. The widths of these peaks correspond to the average mean displacement each particle has relative to this expected spacing, described by the lattice disorder parameter, σ. The nearest neighbor distance can be determined from the maximum of the first peak in the $g(r)$. The insert above shows the difference spectrum, which is the subtraction between the pair correlation function and the hexatic lattice. (**c**) Voronoi tessellations of the particle centroids colored by percent deviation of the expected cell area produced from hexagonal packing. (**d**) Histogram of the probabilities of various percent deviations of the Voronoi cell areas from those produced from hexagonal packing. Taking into account cells with packing of less than 15% of a perfect hexagonal lattice (three blue bars), the film shows ~87% hexagonal packing regions. (**e**) Voronoi tessellations of the particle centroids colored by the normalized bond order parameter. Whiter areas indicate particles that have high angular order in a hexagonal (q_6) symmetry basis. (**f**) Histogram of the probabilities for various values of the bond order parameter. The extracted global bond order parameter is 0.831 $q/q6$ (i.e., 83% of perfect hexagonal symmetry). Expectation value for individual cells (maximum of histogram) is 90% of hexagonal symmetry (**g**) Voronoi tessellation of the particle centroids colored by the number of shared facets (coordination number). (**h**) Histogram of the probabilities for each coordination number. Almost 89% of particles have six nearest neighbors, with only ~5% of matched 5/7 declination pairs.

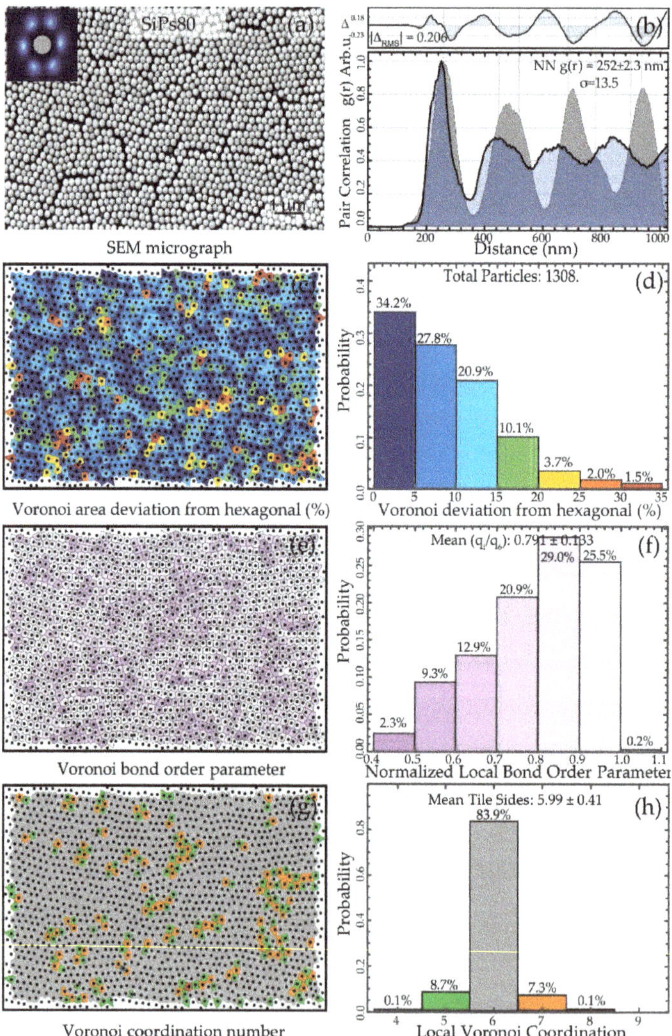

Figure A3. Spatial statistics of sedimented SiPs80. (**a**) SEM image with insert of coordination map or fast Fourier transform (FFT) showing six distinct lobes as expected for hexagonally packed systems. (**b**) pair correlation function for the particle distributions (blue) and that of the matched hexatic lattice (grey). Using the average spacing ($2r_{hex}$) of a hexagonal lattice with the same number density (located at particle positions), the matched hexatic lattice is determined by randomly displacing the particles from their center with a mean distance relative to a normal probability distribution. The widths of these peaks correspond to the average mean displacement each particle has relative to this expected spacing, described by the lattice disorder parameter, σ. The nearest neighbor distance can be determined from the maximum of the first peak in the *g(r)*. The insert above shows the difference spectrum, which is the subtraction between the pair correlation function and the hexatic lattice. (**c**) Voronoi tessellations of the particle centroids colored by percent deviation of the expected cell area produced from hexagonal packing. (**d**) Histogram of the probabilities of various percent deviations of the Voronoi cell areas from those produced from hexagonal packing. Taking into account cells with packing of less than 15% of a perfect hexagonal lattice (three blue bars), the film shows ~83% hexagonal packing regions. (**e**) Voronoi tessellations of the particle centroids colored by the normalized bond order parameter. Whiter areas indicate particles that have high angular order in a hexagonal (q_6) symmetry basis. (**f**) Histogram of the probabilities for various values of the bond order parameter. The extracted global bond order parameter is 0.791 *q/q6* (i.e., 79% of perfect hexagonal symmetry). Expectation value for individual cells (maximum of histogram) is 80–90% of hexagonal symmetry (**g**) Voronoi tessellation of the particle centroids colored by the number of shared facets (coordination number). (**h**) Histogram of the probabilities for each coordination number. Almost 84% of particles have six nearest neighbors, with ~8% of matched 5/7 declination pairs.

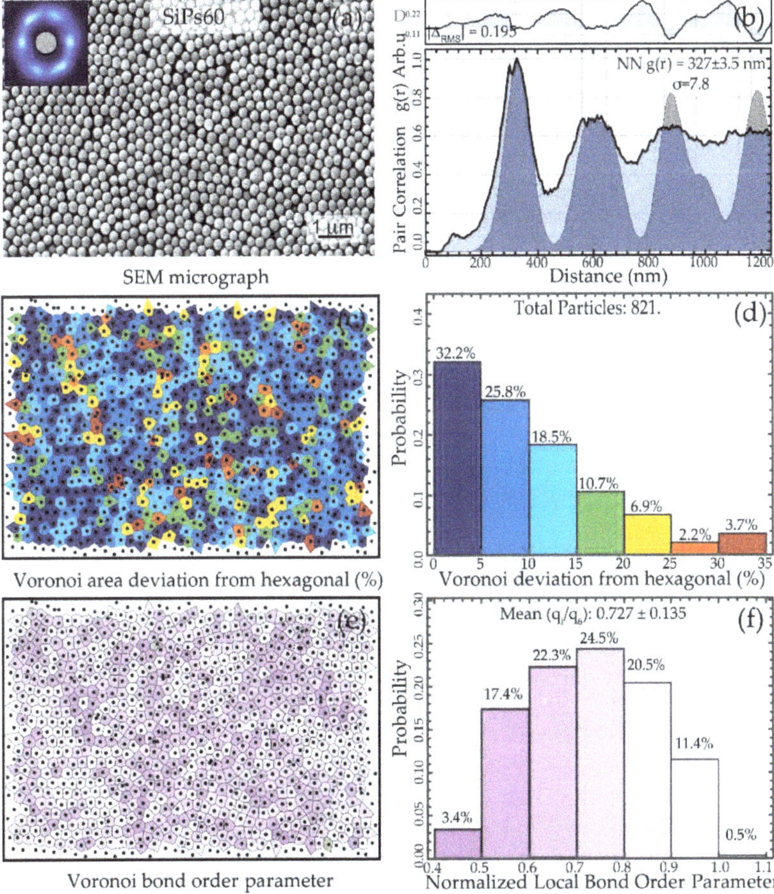

Figure A4. *Cont.*

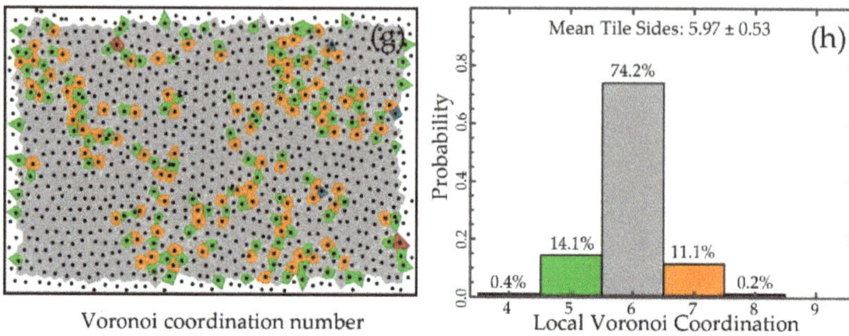

Figure A4. Spatial statistics of sedimented SiPs60. (**a**) SEM image with insert of coordination map or fast Fourier transform (FFT) showing six lobes as expected for hexagonally packed systems. However, the smearing suggests a more powder like disordered configuration rather than sharp lobes. (**b**) pair correlation function for the particle distributions (blue) and that of the matched hexatic lattice (grey). Using the average spacing ($2r_{hex}$) of a hexagonal lattice with the same number density (located at particle positions), the matched hexatic lattice is determined by randomly displacing the particles from their center with a mean distance relative to a normal probability distribution. The widths of these peaks correspond to the average mean displacement each particle has relative to this expected spacing, described by the lattice disorder parameter, σ. The nearest neighbor distance can be determined from the maximum of the first peak in the *g(r)*. The insert above shows the difference spectrum, which is the subtraction between the pair correlation function and the hexatic lattice. (**c**) Voronoi tessellations of the particle centroids colored by percent deviation of the expected cell area produced from hexagonal packing. (**d**) Histogram of the probabilities of various percent deviations of the Voronoi cell areas from those produced from hexagonal packing. Taking into account cells with packing of less than 15% of a perfect hexagonal lattice (three blue bars), the film shows 76.5% hexagonal packing regions. (**e**) Voronoi tessellations of the particle centroids colored by the normalized bond order parameter. Whiter areas indicate particles that have high angular order in a hexagonal (q_6) symmetry basis. (**f**) Histogram of the probabilities for various values of the bond order parameter. The extracted global bond order parameter is 0.727 $q/q6$ (i.e., 73% of perfect hexagonal symmetry). Expectation value for individual cells (maximum of histogram) is 70–80% of hexagonal symmetry (**g**) Voronoi tessellation of the particle centroids colored by the number of shared facets (coordination number). (**h**) Histogram of the probabilities for each coordination number. Almost 74% of particles have six nearest neighbors, with ~13% of matched 5/7 declination pairs.

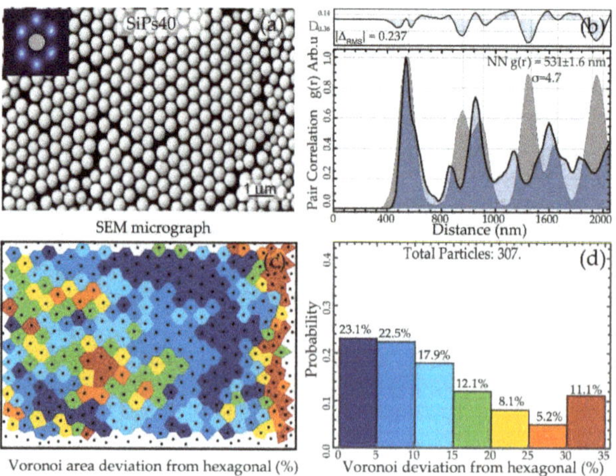

Figure A5. *Cont.*

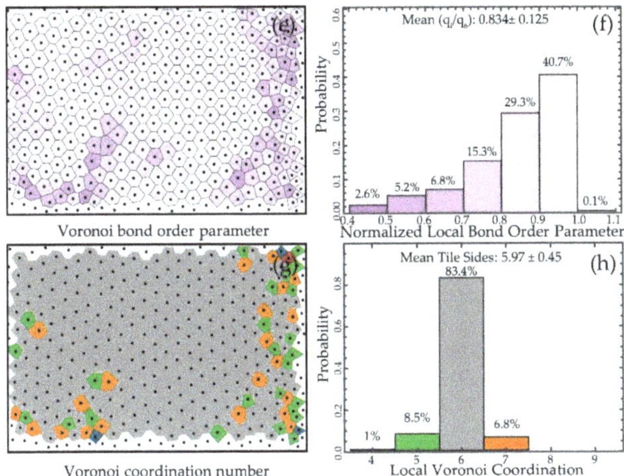

Figure A5. Spatial statistics of sedimented SiPs40. (**a**) SEM image with insert of coordination map or fast Fourier transform (FFT) showing six distinct lobes as expected for hexagonally packed systems (**b**) pair correlation function for the particle distributions (blue) and that of the matched hexatic lattice (grey). Using the average spacing ($2r_{hex}$) of a hexagonal lattice with the same number density (located at particle positions), the matched hexatic lattice is determined by randomly displacing the particles from their center with a mean distance relative to a normal probability distribution. The widths of these peaks correspond to the average mean displacement each particle has relative to this expected spacing, described by the lattice disorder parameter, σ. The nearest neighbor distance can be determined from the maximum of the first peak in the $g(r)$. The insert above shows the difference spectrum, which is the subtraction between the pair correlation function and the hexatic lattice. (**c**) Voronoi tessellations of the particle centroids colored by percent deviation of the expected cell area produced from hexagonal packing. (**d**) Histogram of the probabilities of various percent deviations of the Voronoi cell areas from those produced from hexagonal packing. Taking into account cells with packing of less than 15% of a perfect hexagonal lattice (three blue bars), the film shows 63.5% hexagonal packing regions. This is likely due to the disorganized region visible at the right hand side. (**e**) Voronoi tessellations of the particle centroids colored by the normalized bond order parameter. Whiter areas indicate particles that have high angular order in a hexagonal (q_6) symmetry basis. (**f**) Histogram of the probabilities for various values of the bond order parameter. The extracted global bond order parameter is 0.834 $q/q6$ (i.e., 83% of perfect hexagonal symmetry). Expectation value for individual cells (maximum of histogram) is 90% of hexagonal symmetry (**g**) Voronoi tessellation of the particle centroids colored by the number of shared facets (coordination number). (**h**) Histogram of the probabilities for each coordination number. Over 84% of particles have six nearest neighbors, with ~8% of matched 5/7 declination pairs.

References

1. Fenzl, C.; Hirsch, T.; Wolfbeis, O.S. Photonic crystals for chemical sensing and biosensing. *Angew. Chem. Int. Ed. Engl.* **2014**, *53*, 3318–3335. [CrossRef]

2. Wickham, S.; Poladian, L.; Large, M.; Vukusic, P. Control of iridescence in natural photonic structures: The case of butterfly scales. In *Optical Biomimetics*; Elsevier: Amsterdam, The Netherlands, 2012; pp. 147–176. ISBN 9781845698027.

3. Teyssier, J.; Saenko, S.V.; van der Marel, D.; Milinkovitch, M.C. Photonic crystals cause active colour change in chameleons. *Nat. Commun.* **2015**, *6*, 6368. [CrossRef]

4. Masuda, Y.; Itoh, T.; Koumoto, K. Self-assembly patterning of silica colloidal crystals. *Langmuir* **2005**, *21*, 4478–4481. [CrossRef]

5. Stöber, W.; Fink, A.; Bohn, E. Controlled growth of monodisperse silica spheres in the micron size range. *J. Colloid Interface Sci.* **1968**, *26*, 62–69. [CrossRef]

6. Wong, Y.J.; Zhu, L.; Teo, W.S.; Tan, Y.W.; Yang, Y.; Wang, C.; Chen, H. Revisiting the Stöber method: Inhomogeneity in silica shells. *J. Am. Chem. Soc.* **2011**, *133*, 11422–11425. [CrossRef]

7. Murayama, M.; Howe, J.M.; Hidaka, H.; Takaki, S. Atomic-level observation of disclination dipoles in mechanically milled, nanocrystalline Fe. *Science* **2002**, *295*, 2433–2435. [CrossRef]

8. Bogush, G.H.; Tracy, M.A.; Zukoski, C.F. Preparation of monodisperse silica particles: Control of size and mass fraction. *J. Non-Cryst. Solids* **1988**, *104*, 95–106. [CrossRef]
9. Tan, C.G.; Bowen, B.D.; Epstein, N. Production of monodisperse colloidal silica spheres: Effect of temperature. *J. Colloid Interface Sci.* **1987**, *118*, 290–293. [CrossRef]
10. Malay, O.; Yilgor, I.; Menceloglu, Y.Z. Effects of solvent on TEOS hydrolysis kinetics and silica particle size under basic conditions. *J. Sol-Gel. Sci. Technol.* **2013**, *67*, 351–361. [CrossRef]
11. Danaei, M.; Dehghankhold, M.; Ataei, S.; Hasanzadeh Davarani, F.; Javanmard, R.; Dokhani, A.; Khorasani, S.; Mozafari, M.R. Impact of Particle Size and Polydispersity Index on the Clinical Applications of Lipidic Nanocarrier Systems. *Pharmaceutics* **2018**, *10*, 57. [CrossRef] [PubMed]
12. Mudalige, T.; Qu, H.; van Haute, D.; Ansar, S.M.; Paredes, A.; Ingle, T. Characterization of Nanomaterials. In *Nanomaterials for Food Applications*; Elsevier: Amsterdam, The Netherlands, 2019; pp. 313–353. ISBN 9780128141304.
13. Park, T.J.; Lee, S.-K.; Yoo, S.M.; Yang, S.-M.; Lee, S.Y. Development of reflective biosensor using fabrication of functionalized photonic nanocrystals. *J. Nanosci. Nanotechnol.* **2011**, *11*, 632–637. [CrossRef]
14. Wu, Z.; Hu, X.; Tao, C.; Li, Y.; Liu, J.; Yang, C.; Shen, D.; Li, G. Direct and label-free detection of cholic acid based on molecularly imprinted photonic hydrogels. *J. Mater. Chem.* **2008**, *18*, 5452. [CrossRef]
15. Zhang, Y.; Jin, Z.; Zeng, Q.; Huang, Y.; Gu, H.; He, J.; Liu, Y.; Chen, S.; Sun, H.; Lai, J. Visual test for the presence of the illegal additive ethyl anthranilate by using a photonic crystal test strip. *Mikrochim. Acta* **2019**, *186*, 685. [CrossRef]
16. Gao, W.; Rigout, M.; Owens, H. Self-assembly of silica colloidal crystal thin films with tuneable structural colours over a wide visible spectrum. *Appl. Surf. Sci.* **2016**, *380*, 12–15. [CrossRef]
17. Koay, N.; Burgess, I.B.; Kay, T.M.; Nerger, B.A.; Miles-Rossouw, M.; Shirman, T.; Vu, T.L.; England, G.; Phillips, K.R.; Utech, S.; et al. Hierarchical structural control of visual properties in self-assembled photonic-plasmonic pigments. *Opt. Express* **2014**, *22*, 27750–27768. [CrossRef]
18. Shen, Z.; Shi, L.; You, B.; Wu, L.; Zhao, D. Large-scale fabrication of three-dimensional ordered polymer films with strong structure colors and robust mechanical properties. *J. Mater. Chem.* **2012**, *22*, 8069. [CrossRef]
19. Li, H.; Tang, Q.; Cai, F.; Hu, X.; Lu, H.; Yan, Y.; Hong, W.; Zhao, B. Optimized photonic crystal structure for DSSC. *Sol. Energy* **2012**, *86*, 3430–3437. [CrossRef]
20. Liu, W.; Ma, H.; Walsh, A. Advance in photonic crystal solar cells. *Renew. Sustain. Energy Rev.* **2019**, *116*, 109436. [CrossRef]
21. Yablonovitch, E. Inhibited spontaneous emission in solid-state physics and electronics. *Phys. Rev. Lett.* **1987**, *58*, 2059–2062. [CrossRef]
22. van Dommelen, R.; Fanzio, P.; Sasso, L. Surface self-assembly of colloidal crystals for micro- and nano-patterning. *Adv. Colloid Interface Sci.* **2018**, *251*, 97–114. [CrossRef]
23. Yeo, S.J.; Choi, G.H.; Yoo, P.J. Multiscale-architectured functional membranes utilizing inverse opal structures. *J. Mater. Chem. A* **2017**, *5*, 17111–17134. [CrossRef]
24. Yablonovitch, E.; Gmitter, T.J.; Leung, K.M. Photonic band structure: The face-centered-cubic case employing nonspherical atoms. *Phys. Rev. Lett.* **1991**, *67*, 2295–2298. [CrossRef]
25. Bumstead, M.; Liang, K.; Hanta, G.; Hui, L.S.; Turak, A. disLocate: Tools to rapidly quantify local intermolecular structure to assess two-dimensional order in self-assembled systems. *Sci. Rep.* **2018**, *8*, 1554. [CrossRef]
26. Gao, W.; Rigout, M.; Owens, H. Facile control of silica nanoparticles using a novel solvent varying method for the fabrication of artificial opal photonic crystals. *J. Nanopart. Res.* **2016**, *18*, 387. [CrossRef]
27. Agarwal, U.; Escobedo, F.A. Mesophase behaviour of polyhedral particles. *Nat. Mater.* **2011**, *10*, 230–235. [CrossRef] [PubMed]
28. Timmermans, J. Plastic crystals: A historical review. *J. Phys. Chem. Solids* **1961**, *18*, 1–8. [CrossRef]
29. Bernard, E.P.; Krauth, W. Two-step melting in two dimensions: First-order liquid-hexatic transition. *Phys. Rev. Lett.* **2011**, *107*, 155704. [CrossRef]
30. Williamson, J.J.; Evans, R.M.L. Measuring local volume fraction, long-wavelength correlations, and fractionation in a phase-separating polydisperse fluid. *J. Chem. Phys.* **2014**, *141*, 164901. [CrossRef] [PubMed]
31. Okabe, A.; Boots, B.; Sugihara, K.; Chiu, S.N.; Kendall, D.G. *Spatial Tessellations*; John Wiley & Sons, Inc.: Hoboken, NJ, USA, 2000; ISBN 9780470317013.
32. van Anders, G.; Ahmed, N.K.; Smith, R.; Engel, M.; Glotzer, S.C. Entropically patchy particles: Engineering valence through shape entropy. *ACS Nano* **2014**, *8*, 931–940. [CrossRef] [PubMed]
33. Agarwal, U.; Escobedo, F.A. Effect of quenched size polydispersity on the ordering transitions of hard polyhedral particles. *J. Chem. Phys.* **2012**, *137*, 24905. [CrossRef] [PubMed]
34. Damasceno, P.F.; Engel, M.; Glotzer, S.C. Crystalline assemblies and densest packings of a family of truncated tetrahedra and the role of directional entropic forces. *ACS Nano* **2012**, *6*, 609–614. [CrossRef]
35. Kirkwood, J.G.; Boggs, E.M. The Radial Distribution Function in Liquids. *J. Chem. Phys.* **1942**, *10*, 394–402. [CrossRef]
36. Steinhardt, P.J.; Nelson, D.R.; Ronchetti, M. Bond-orientational order in liquids and glasses. *Phys. Rev. B* **1983**, *28*, 784–805. [CrossRef]
37. Magonov, S.N. *Surface Analysis with STM and AFM: Experimental and Theoretical Aspects of Image Analysis*; VCH: Weinheim, Germany; New York, NY, USA, 2008; ISBN 9783527615100.

38. Bharati, M.H.; Liu, J.; MacGregor, J.F. Image texture analysis: Methods and comparisons. *Chemom. Intell. Lab. Syst.* **2004**, *72*, 57–71. [CrossRef]
39. Rankin, D.W.H.; Morrison, C.A.; Mitzel, N.W. *Structural Methods in Molecular Inorganic Chemistry*; Wiley: Chichester, West Sussex, UK, 2013, ISBN 9781118462881.
40. Ladd, M.; Palmer, R. *Structure Determination by X-ray Crystallography*; Springer US: Boston, MA, USA, 2013, ISBN 978-1-4614-3956-1.
41. Barón, M. Definitions of basic terms relating to low-molar-mass and polymer liquid crystals (IUPAC Recommendations 2001). *Pure Appl. Chem.* **2001**, *73*, 845–895. [CrossRef]
42. Lobmeyer, D.M.; Biswal, S.L. Grain boundary dynamics driven by magnetically induced circulation at the void interface of 2D colloidal crystals. *Sci. Adv.* **2022**, *8*, eabn5715. [CrossRef]

 *materials*

Article

Photoaligned Liquid Crystal Devices with Switchable Hexagonal Diffraction Patterns

Inge Nys *, Brecht Berteloot and Kristiaan Neyts *

Liquid Crystals & Photonics Group, Department of Electronics and Information Systems, Ghent University, Technologiepark-Zwijnaarde 126, 9052 Ghent, Belgium; brecht.berteloot@ugent.be
* Correspondence: inge.nys@ugent.be (I.N.); kristiaan.neyts@ugent.be (K.N.)

Abstract: Highly efficient optical diffraction can be realized with the help of micrometer-thin liquid crystal (LC) layers with a periodic modulation of the director orientation. Electrical tunability is easily accessible due to the strong stimuli-responsiveness in the LC phase. By using well-designed photoalignment patterns at the surfaces, we experimentally stabilize two dimensional periodic LC configurations with switchable hexagonal diffraction patterns. The alignment direction follows a one-dimensional periodic rotation at both substrates, but with a 60° or 120° rotation between both grating vectors. The resulting LC configuration is studied with the help of polarizing optical microscopy images and the diffraction properties are measured as a function of the voltage. The intricate bulk director configuration is revealed with the help of finite element Q-tensor simulations. Twist conflicts induced by the surface anchoring are resolved by introducing regions with an out-of-plane tilt in the bulk. This avoids the need for singular disclinations in the structures and gives rise to voltage induced tuning without hysteretic behavior.

Keywords: nematic liquid crystal; 2D periodic structures; hexagonal diffraction patterns; photoalignment; out-of-plane reorientation; flat optical elements

Citation: Nys, I.; Berteloot, B.; Neyts, K. Photoaligned Liquid Crystal Devices with Switchable Hexagonal Diffraction Patterns. *Materials* **2022**, *15*, 2453. https://doi.org/10.3390/ma15072453

Academic Editors: Ivan V. Timofeev and Wei Lee

Received: 21 February 2022
Accepted: 16 March 2022
Published: 26 March 2022

Publisher's Note: MDPI stays neutral with regard to jurisdictional claims in published maps and institutional affiliations.

1. Introduction

Thanks to the remarkable combination of fluidity and long-range ordering in the liquid crystal (LC) phase, LC-related research has continued to attract attention over many decades [1–4]. The long-range ordering of the anisotropic molecules in the LC phase leads to macroscopic anisotropy, with directional dependence of the optical, electrical, magnetic, mechanical, etc. properties. The fluidity, on the other hand, gives rise to strong stimuli-responsiveness, resulting in easily accessible tunability with the help of electric fields, heating, illumination, etc. During the last few decades, high-resolution photoalignment techniques became available to pattern the anchoring of the LC at the confining substrates [4–6]. This offers the possibility to steer the self-organization of complex 3D LC configurations in the bulk and in this way engineer devices with desired functionality. In many applications, the LC is used for its electro-optical functionality. The most well-known example is the LC display (LCD), but nowadays LC is also being used to an increasing extent in beam splitters, lenses, diffraction gratings, lasers, beam shaping elements, spatial light modulators, etc. [4,7–17]. Photoalignment is an important enabling technology to design these kinds of flat optical components, where efficient light manipulation is obtained in micrometer-thin LC layers. Molecules in the photoalignment layer typically tend to orient perpendicularly to the polarization direction of the illuminating light, and a well-designed alignment pattern can be imposed by using structured illumination. To do so, different illumination methods have been developed: interference illumination, direct write illumination, plasmonic patterning, structured illumination with the help of a spatial light modulator, etc. [18–22]. The cell can be assembled before or after doing the illumination and in this way different alignment patterns can be imposed at the top and bottom substrate. This offers additional possibilities to design LC components with complex director configurations and advanced optical functionalities.

We here impose a 1D periodically rotating alignment pattern at the top and bottom substrate, with a deviating rotation direction at both substrates (Figure 1). The cells are filled with nematic liquid crystal (NLC) with a positive dielectric anisotropy. The complex alignment configurations at the interfaces give rise to intricate periodic director arrangements in the bulk and electrically tunable 2D diffraction patterns. We investigate the effect of different rotation angles ψ ($60°$, $90°$, $120°$) between the top and bottom alignment pattern on the bulk director configuration and on the diffraction characteristics, by comparing experiments with numerical simulations.

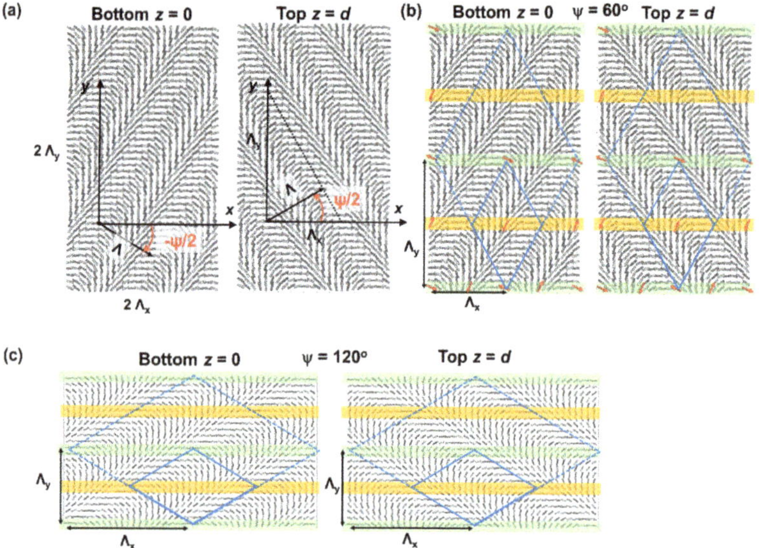

Figure 1. Alignment configuration at the bottom ($z = 0$) and top ($z = d = 5.5\ \mu\mathrm{m}$) substrate for $\psi = 60°$ (**a,b**) $\psi = 120°$ (**c**). One simulated unit cell is shown. Periodic boundary conditions are applied at the edges with a fixed x- or y-coordinate. (**b,c**) Indication of regions with parallel (green) and anti-parallel (yellow) director alignment at the top and bottom substrate when a vector representation of the director is used. The unit cell for the bulk LC director configuration is shown in dashed lines, together with the unit cell for the alignment pattern in full lines.

A similar alignment configuration has been investigated before, but only for crossed assembly of both substrates ($\psi = 90°$) [22–27] or anti-parallel assembly ($\psi = 180°$) [28,29]. Somewhat different from the other alignment configurations, $\psi = 180°$ gives rise to a 1D configuration. Regarding the 2D configurations, Crawford et al. in 2005 and Provenzano et al. in 2007 were the first to study the crossed alignment configuration ($\psi = 90°$) experimentally (in static conditions) [22,23]. Wang et al. used this alignment configuration in 2017 in combination with sliding substrate assembly [24]. We were the first to investigate the director configuration in detail for the crossed assembly of rotating planar alignment patterns [25–27]. We demonstrated that surface-induced twist conflicts can be elegantly resolved by introducing a region with vertical director orientation in the bulk, a possibility that was overlooked before. Although the alignment pattern induces planar anchoring at the substrates, the director becomes vertical in a region in the bulk to avoid the creation of disclination lines and minimize the free energy. In order to do so, symmetry breaking takes place, leading to a LC superstructure with a larger period (unit cell $2\Lambda \times 2\Lambda$) than defined by the boundary conditions (unit cell $\Lambda \times \Lambda$). In this article, we demonstrate that this concept can be generalized to other (non-$90°$) rotation angles between the alignment patterns at both substrates. The self-organization of disclination-free LC configurations with local regions with vertical director orientation is demonstrated experimentally and with the help of numerical simulations. The regions with close-to-vertical mid-plane director orientation are approximately lines along a bisector between the axes of director variation on both substrates. The voltage-induced tunability of the director configuration and the diffraction characteristics are also discussed.

2. Materials and Methods

2.1. Sample Preparation

To prepare the cells, two ITO-coated glass substrates are covered with a photoalignment layer and illuminated with the help of an SLM based illumination setup. As a photoalignment layer, we use Brilliant Yellow (BY, Sigma-Aldrich, St. Louis, MO, USA, 0.2 wt %) dissolved in dimethylformamide (Sigma-Aldrich). After cleaning, the substrates are treated with an ozone-plasma and the photoalignment layer is spincoated (3000 rpm, 30 s). The substrates are subsequently dried on a hotplate for 5 min at 90 °C and the photoalignment procedure is followed before the substrates are glued together. Spherical spacers with 5.5 μm diameter are mixed in the glue (NOA 68) and only the glue edges are illuminated by UV light for curing, while the rest of the cell is protected by aluminum foil. After gluing the substrates together, the cell is filled with nematic LC E7 at 80 °C (above the transition temperature) and cooled down to room temperature. For the photoalignment procedure, a blue laser (Cobolt Twist, 200 mW, $\lambda = 457$ nm) is combined with a spatial light modulator (Holoeye Pluto 2) with a resolution of 1920 by 1080 pixels and a pixel pitch of 8 μm. With the help of the necessary quarter-wave plates, the pattern that is displayed on the SLM defines the (linearly polarized) polarization pattern projected on the substrate. The setup itself is discussed in more detail in previous articles [16,20].

After illumination, the substrates are assembled so that there is a ψ rotation between the 1D rotating alignment pattern at the top and bottom substrate. The azimuthal angle φ w.r.t. the x-axis is varying at the substrates as

$$\varphi(r) = \varphi_0 - \mathbf{K}.r$$
$$\mathbf{K}_{t,b} = \frac{\pi}{\Lambda}\cos\frac{\psi}{2}\,\mathbf{1}_x \; \pm \; \frac{\pi}{\Lambda}\sin\frac{\psi}{2}\,\mathbf{1}_y \tag{1}$$

with $\mathbf{K}_t$ and $\mathbf{K}_b$ respectively being the grating vector at the top and bottom substrate. The alignment period Λ was equal to 10 μm. To obtain alignment configurations with different ψ in the same cell, the alignment patterns imposed by the SLM are rotated over the appropriate angle for different illuminated areas. In our convention, the alignment patterns are rotated over $\pm\psi/2$ around the z-axis for the top and bottom substrate (Figure 1). The bisectors between the rotated axes are along the x- and the y-axis. With these definitions, we find linear regions parallel to the x-axis in which the top and bottom alignment are parallel, indicated by yellow and green in Figure 1. The boundary conditions are periodic in the y-direction with period Λ_y

$$\Lambda_y = \frac{\Lambda}{\sin\left(\frac{\psi}{2}\right)} \tag{2}$$

and in the x-direction with period Λ_x

$$\Lambda_x = \frac{\Lambda}{\cos\left(\frac{\psi}{2}\right)}. \tag{3}$$

2.2. FE Q-Tensor Simulations and Simulations of Optical Microscope Images

To simulate the 3D director configuration in the bulk of the cell, finite element Q-tensor simulations are used [30,31]. The representation of the LC in terms of the Q-tensor—combining information about the LC order and orientation—can simulate configuration with and without singular disclinations. The simulation program is looking for a stationary solution that minimized the Landau De Gennes free energy, being the sum of the elastic distortion energy, electric energy (when a voltage is applied), surface energy, and thermotropic (or Landau) energy [32,33]. The energy contributions are expressed in terms of the Q-tensor and its spatial derivatives and material parameter of the LC E7 are taken into account (K_{11} = 11.1 pN, K_{22} = 6.5 pN, K_{33} = 17.1 pN, ε_{perp} = 5.2, ε_{para} = 19). The values of the bulk thermotropic coefficients A, B, and C (A = -174 N/m^2, B = -2120 N/m^2, C = 1740 N/m^2) are based on the experimentally measured values for 5CB at a reduced temperature of -2 °C [34]. Smaller values for these constants are used to increase the natural length scale of variations in the order parameter and to obtain faster numerical convergence [35]. A tetrahedral mesh is used to describe the volume of the LC and a voltage can be applied between the $z = 0$ and $z = d$ coordinate (uniform electrodes at the substrates). Periodic boundary conditions are used at the edges of the simulation domain for constant x- or y-coordinate, and fixed anchoring is imposed at the substrates $z = 0$ and $z = d = 5.5$ μm. Since we can only impose periodic boundary conditions for a rectangular unit cell, with edges for a fixed x- and y-coordinate, we simulated a rectangular unit cell as shown in Figure 1.

Based on the simulation results for the director configuration, optical simulations for the microscopy images are performed with the help of the open source software Nemaktis (https://github.com/warthan07/Nemaktis (accessed on 25 March 2022)). This software implements part of the generalized beam propagation method for birefringent media described by Poy et al. and can include the focusing optics of the microscopy and the spectrum of the illuminant [36]. Simulations are done for 15 different wavelengths equally distributed over the visible wavelength range and the spectrum of a CIE. An illuminant is used to weight the different contributions and obtain a color image.

3. Experimental Results

3.1. Microscopy Images

Experimentally measured polarizing optical microscopy (POM) images are shown in Figures 2 and 3 for gratings with a different rotation angle between the top and bottom substrate. Results without applied voltage are shown in Figure 2 (for $\psi = 60°, 90°, 120°$) while measurements for different applied voltages are shown in Figure 3 (for $\psi = 60°$ and $120°$). The frequency of the applied electric field is 1 kHz. The simulated unit cell is indicated in the microscopy images. In all cases, a disclination-free director configuration is observed in the bulk of the cell. The structure smoothly varies when a voltage is applied and no hysteresis behavior occurs.

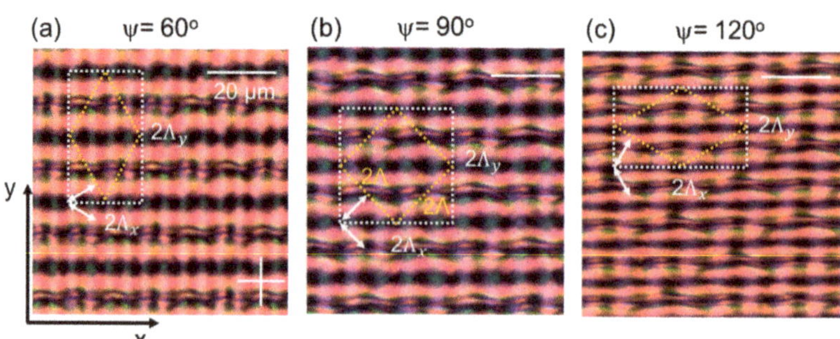

Figure 2. Experimental polarization optical microscopy images for gratings with $\psi = 60°, 90°, 120°$, from left to right (**a–c**). The polarizer and analyzer are crossed and oriented along the x- and y-axes respectively. The grating vectors at the top and bottom substrate are indicated with white arrows. The simulated unit cell, with dimension $2\Lambda_x \times 2\Lambda_y$, is shown in a white dotted rectangle while the unit cell for the bulk director configuration is indicated in yellow.

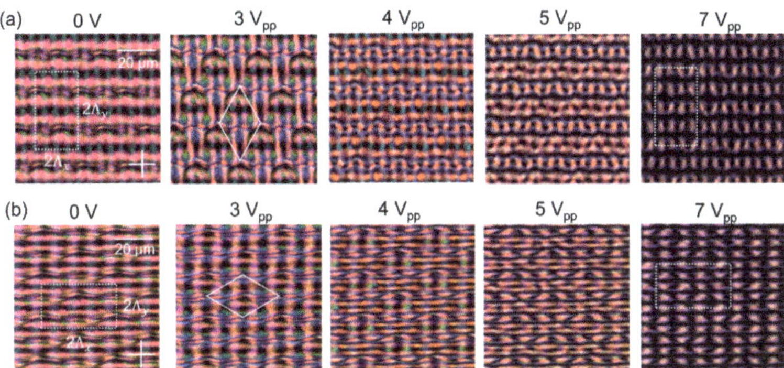

Figure 3. Experimental polarization optical microscopy images for gratings with $\psi = 60°$ (**a**) $\psi = 120°$ (**b**). The applied voltage is increasing from left (0 V) to right (7 V_{pp}). The polarizer and analyzer are crossed and oriented along the x- and y-axes respectively. The simulated unit cell, with dimension $2\Lambda_x \times 2\Lambda_y$, is shown in a white dotted rectangle (for 0 V and 7 V_{pp}) and the bulk unit cell is shown in a white diamond (for 3 V_{pp}).

The microscopy images immediately show that similarities exist between the alignment configurations with different rotation angles ψ. Although the dimensions of the unit cell are stretched for different alignment configurations, horizontal bands of darker and brighter appearance can be clearly recognized in all images (Figure 2). The spacing along the y-direction becomes smaller for increasing ψ. The dark bands correspond to the regions with (anti-)parallel alignment at the top and bottom substrate as indicated in Figure 1. Within the simulated area with $2\Lambda_x \times 2\Lambda_y$ dimension, four brighter and four darker horizontal bands appear. Without applied voltage, a clear distinction can be seen between two types of dark bands, occurring alternatingly. One of them is wider and consists of dark and slightly more bright regions next to each other (along x) with little variation. The other type is shifted along the y-direction over a distance $\Lambda_y/2$ and contains a thin dark line. The different bright horizontal bands (shifted along the y-direction over a distance $\Lambda_y/2$) are less easily distinguished at 0 V. For slightly increased voltages the distinction becomes more clear and the unit cell can be better recognized (Figure 3). The exact optical appearance and the observed colors depend on the alignment configuration and cell thickness.

At sufficiently high voltages (>6 V_{pp}), the symmetry in the POM images increases and a grid of bright points is observed (Figure 3). The director is oriented close-to-vertically in the bulk and appreciable deviations from the vertical direction are only present close to the photoaligned interfaces. Between crossed polarizers, no light is transmitted when the azimuthal anchoring at the top and bottom substrate is in perpendicular directions. Bright areas in the POM images correspond to regions where the director at the top and bottom substrate is parallel and under an angle (of ~45°) with respect to the crossed polarizers. For the tested alignment configuration, the lines with parallel top and bottom alignment are running parallel to the x-axis, with a spacing along the y-direction equal to $\Lambda_y/2 = \Lambda/(2 \sin(\psi/2))$. The spacing between bright points along the x-axis is $\Lambda_x/2 = \Lambda/(2 \cos(\psi/2))$. For $\psi = 90°$, as studied before [25], the structure becomes centro-symmetric at high voltages with a periodic pattern along the x- and y-axis with a $\Lambda/\sqrt{2}$ periodicity in both directions).

3.2. Diffraction Measurements

The intricate 2D alignment configuration, composed of 1D periodically rotating alignment patterns at both substrates, gives rise to interesting diffraction behavior. The working principle can be understood in terms of the geometric phase or so called Pancharatnam–Berry phase of light [37,38]. Very efficient geometric phase gratings, working in transmission or reflection, have been demonstrated before [11–14]. The focus is often on 1D gratings or lens configurations, but here we investigate 2D gratings with different symmetry. The voltage-dependent diffraction characteristic was measured for an alignment configuration with $\psi = 60°$ and $120°$ and important features are explained. These particular rotation angles between the alignment patterns at the top and bottom substrate give rise to diffraction patterns with hexagonal symmetry.

Experimentally observed diffraction patterns for different applied voltages are shown in Figure 4 for $\psi = 60°$ (a,b) and $\psi = 120°$ (c). The incident light from a helium–neon laser ($\lambda = 633$ nm) was circularly polarized in (a) and linearly polarized along the y-axis in (b) and (c). A polarizer was used to improve the degree of linear polarization and an additional quarter-wave plate was inserted for the experiments in Figure 4a. A circular area with a diameter of approximately 1 mm was illuminated and the incident beam power was 1120 µW. The light was captured on a screen behind the LC cell (in transmission) and images of the screen were taken with a smartphone camera. The cell was mounted so that the incident light beam first hits the top substrate and leaves the cell through the bottom substrate. For these geometric phase gratings, the diffraction pattern is asymmetric (with respect to inversion of k_x and k_y) for circularly polarized incident light. For incident light with the opposite circular polarization the same results are obtained but rotated over 180 degrees. For a plane monochromatic wave incident on a purely periodic structure, the projection of the wave vector on the xy-plane for each diffracted beam is given by

$$\mathbf{k}_{xy}(i,j) = i\,2\mathbf{K}_t + j\,2\mathbf{K}_b \tag{4}$$

with $\mathbf{K}_b$ and $\mathbf{K}_t$ being the vectors indicating the rotation of the photoalignment at the bottom and top substrate as defined in Equation (1). The indices i and j indicate the order of diffraction. Diffraction order $(-1, 0)$ corresponds with diffraction in the -1st order associated with the top substrate, while $(0, -1)$ corresponds to -1st order diffraction associated with the bottom substrate (Figure 4d). Because the period of the bulk configuration is larger than the period at the substrate, indices with halve integers are also found in the diffraction pattern (Figure 4).

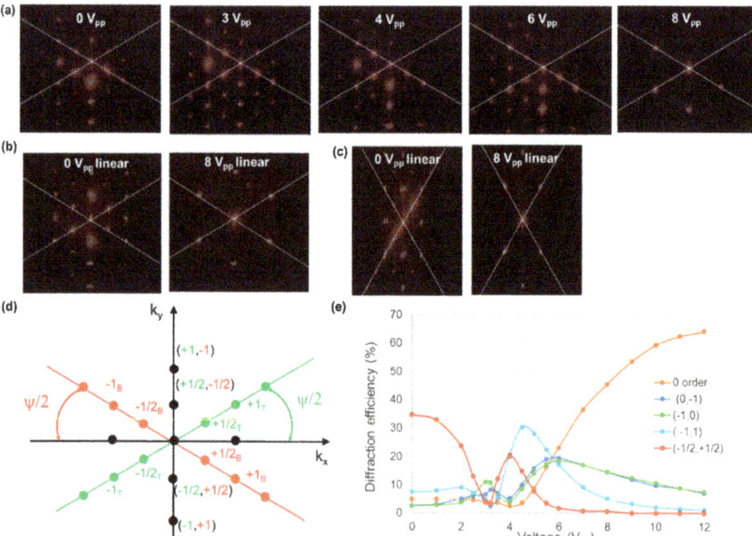

Figure 4. Experimental diffraction images at different voltages for a grating with $\psi = 60°$ (**a,b**) and $\psi = 120°$ (**c**). The incident light was circularly polarized in (**a**) and linearly polarized in (**b,c**). A schematic representation of the diffraction orders for $\psi = 60°$ is shown in (**d**). The measured power in four different diffraction orders is shown in figure (**e**) for $\psi = 60°$ and circularly polarized incident light (corresponding to experiment (**a**)).

The symmetry of the alignment configuration can clearly be recognized in the diffraction patterns (Figure 4). The distribution of light into the different orders is rather complex and non-monotonous behavior is observed as a function of the voltage since the cell is relatively thick (Δnd$/\lambda$ = 0.2 × 5.5 μm/0.633 μm >> 1/2). For the experiments shown in Figure 4a, the experimentally measured power into the most important diffraction orders is summarized in Figure 4e. The diffraction order that is strongest at 0 V is diffracted towards the negative k_y-axis. This order (denoted $(-1/2, 1/2)$ in Figure 4d) first decreases for increasing voltages, but shows an additional maximum around 4 V_{pp} before going to zero at high voltages. The zero-order transmission is small at low voltages but becomes dominant at high voltages.

4. Simulation Results

To understand how the LC director is oriented in the bulk of the layer, numerical simulations are performed. The simulated area is a rectangle with dimension $2\Lambda_x * 2\Lambda_y = 8 \times \Lambda^2/\sin\psi$ (Figures 1 and 2). This is two times larger than the unit cell of the bulk LC configuration (Figure 2), since our simulation tool can only impose periodic boundary conditions in a rectangular volume with bounding plates parallel to the coordinate planes. In the rectangular unit cell with basis $2\Lambda_x$ and height $2\Lambda_y$ periodic boundary conditions apply (Figure 1).

In Figures 5 and 6, the simulation results are summarized for different applied voltages for ψ = 60° and ψ = 120° respectively. The details of the director configuration depend on the parameters Λ, d, and ψ, but some general observations can be made. When observing the mid-plane cross-section (Figures 5a and 6a) it is clear that alternating regions with high and low tilt angle appear along the y-axis with period Λ_y. In the simulated area, two line-like regions appear with close-to-vertical mid-plane director orientation at 0 V for $y = \Lambda_y/2$ and $y = 3\Lambda_y/2$. This corresponds to positions in the alignment configuration where the director at the top and bottom substrate is anti-parallel as shown in Figure 1. In between, for $y \approx 0$ or $y \approx \Lambda_y$, the alignment at the top and bottom substrate is parallel (Figure 1) and the director in the bulk remains roughly parallel to the substrates (Figures 5 and 6), with the same azimuthal angle as defined at the substrates. Some deviation from the planar alignment in the bulk is observed in these regions, with modulation of the tilt-angle along the x-direction (Figures 5d and 6d). This phenomenon was also observed before in cells with ψ = 90° and the deviations from the planar orientation are in this case related to the Λ/d ratio. Decreasing Λ/d leads to increasing deviations from the horizontal mid-plane alignment and an increased asymmetry in the unit cell [25–27]. The fact that these line-regions at $y \approx 0$ or $y \approx \Lambda_y$ appear rather dark in the POM images at 0 V (Figures 2, 5b, and 6b) is linked to the retardation in the LC layer. By also looking at the results for other voltages, the regions with azimuthal orientation along one of the polarizers (dark) or under ~45° angle w.r.t. the polarizers (bright) can be better recognized.

(a) Mid-plane configuration z = d/2 = 2.75 μm

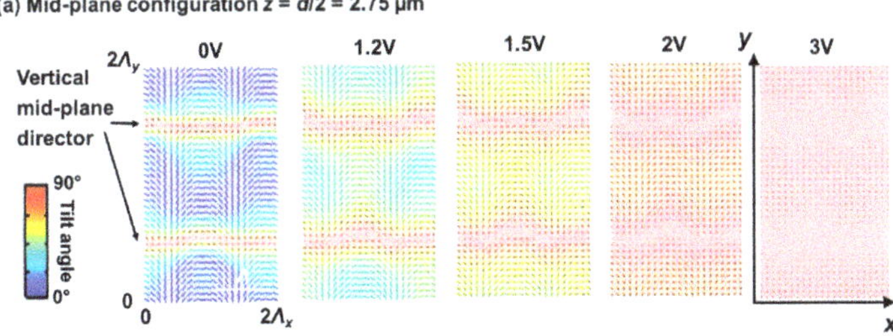

Figure 5. *Cont.*

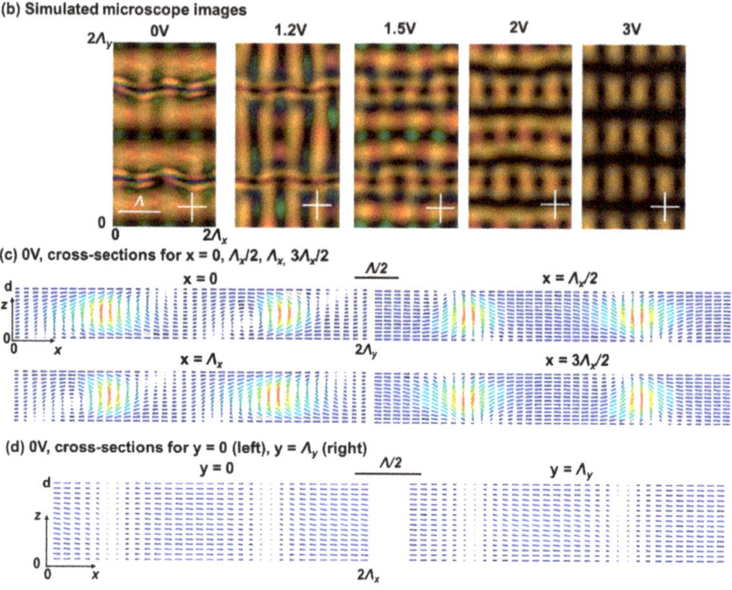

Figure 5. Simulated director configurations (**a,c,d**) and transmission images (**b**) for ψ = 60°. The mid-plane ($z = d/2 = 2.75$ μm) director configuration is shown in (**a**) for different applied voltages (root mean square values). Cross-sections for constant x- and y-coordinate are shown in (**c,d**) at 0 V. The color of the director indicates the tilt angle with respect to the xy-plane. The alignment period Λ is 10 μm.

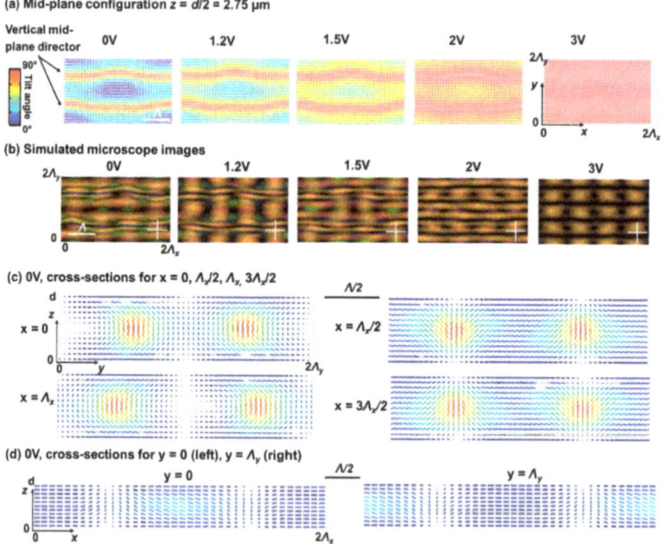

Figure 6. Simulated director configurations (**a,c,d**) and transmission images (**b**) for ψ = 120°. The mid-plane ($z = d/2 = 2.75$ μm) director configuration is shown in (**a**) for different applied voltages (root mean square values). Cross-sections for constant x- and y-coordinates are shown in (**c,d**) at 0 V. The color of the director indicates the tilt angle with respect to the xy-plane. The alignment period Λ is 10 μm.

5. Discussion

In general, very good agreement is observed between experimental POM images (Figures 2 and 3) and simulated microscope images (Figures 5b and 6b). This agreement proves that the complex bulk director configuration is well-understood. Minor changes between both can be related to small changes in the exact cell thickness, illumination conditions, imperfections in the SLM photoalignment illumination, and focusing effects in the microscope. Moreover, strong anchoring is assumed in the simulations while finite anchoring strength will slightly influence the optical properties.

For the studied alignment configurations, local out-of-plane reorientation of the director can resolve the twist-conflicts induced by the substrates. In order to do so, the unit cell for the bulk director configuration needs to be enlarged with respect to the alignment configuration at the substrates. The unit cell for the alignment configuration has area $\frac{1}{2} \times \Lambda_x \times \Lambda_y$ (Figure 1) while the unit cell of the bulk structure has a parallelogram shape with area $2 \times \Lambda_x \times \Lambda_y$. Neighboring horizontal lines spaced over $\Lambda_y/2$ in the y-direction, with equal top and bottom alignment, correspond with another bulk director orientation. This can be intuitively visualized by representing the alignment pattern with the help of a vector, having a head and tail (Figure 1). Although the alignment pattern itself imposes planar anchoring without pretilt, out-of-plane reorientation is present in the bulk. Therefore, regions with parallel or anti-parallel surface orientation of the vector are non-equivalent. In the bulk, these two-line regions (e.g., at $y = 0$ and $y = \Lambda_y/2$) behave differently, with an alteration of line regions with approximately parallel director and line regions with close-to-homeotropic orientation in the mid-plane. These observations are similar to what we observed before in cells with crossed assembly ($\psi = 90°$) of the alignment patterns [25–27]. Although the details of the director configurations are obviously different, the general concept behind the structural formation in the bulk can be generalized. Local out-of-plane director orientation in the bulk can resolve twist conflicts at the aligning substrates for all tested rotation angles ψ.

The fact that symmetry breaking takes place to form the bulk LC configuration is inherently linked to the existence of multiple equivalent solutions within the unit cell. In an experimental device, different shifted domains can exist. A disclination line is present between two shifted domains, as can be seen in Figure 7. These borders between the domains break the periodicity and lead to some background of light between the integer (or half integer) diffraction orders. In general, the fact that the boundary conditions can lead to multiple equivalent bulk structures offers potential for multi-stable devices that can be switched electrically or optically with low power consumption [28,39,40].

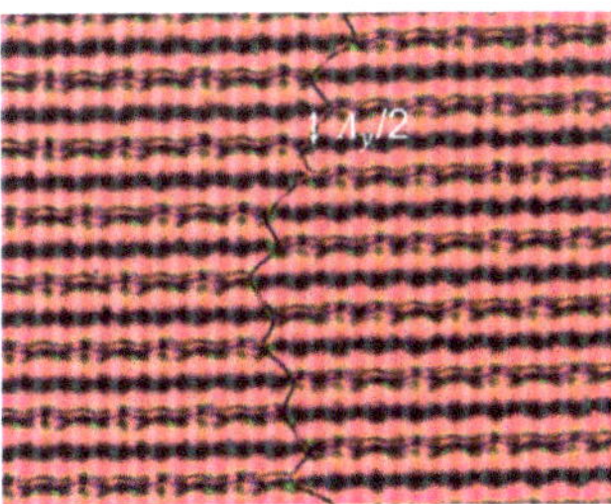

Figure 7. POM image showing the existence of two shifted domains separated by a disclination line. $\psi = 60°$ in this example.

It is worth noting that for a similar alignment configuration with $\psi = 90°$, Wang et al. reported the formation of a web of twist-disclination lines in the bulk when using sliding substrate assembly [24]. To avoid the need for these kinds of singular disclinations in the structure, regions with a close-to-vertical mid-plane director orientation are necessary. In our experiments with fixed substrates (with $d = 5.5\ \mu m$ and $\Lambda = 10\ \mu m$) we did not observe a web of disclination lines for any of the studied alignment configurations ($\psi = 60°, 90°,$

120°). A defect-free director configuration, with regions with vertical director orientation, is naturally favored after going through a voltage cycle (shortly applying a high voltage). In our experiments, this disclination-free configuration was stable for months and this structure was also formed immediately after cooling from the isotropic phase without applied voltage. This is different from the results reported by Honma et al. in a cell with large alignment period Λ = 80 μm and rotation angle ψ = 180° (1D configuration). In this case, bistable switching between different configurations (with and without singular disclinations) was observed [28].

The patterned structures with ψ = 60° or 120° give rise to hexagonal diffraction patterns with easy electrical tunability. The application of a small voltage between the uniform electrodes at the top and bottom substrate can redistribute power into different orders. Numerical simulations are necessary to predict the diffraction efficiency and its voltage dependence (see [25] for ψ = 90°), but the general concepts can be understood from a theoretical basis as outlined below.

The behavior of 1D gratings with the same periodically rotating alignment pattern at the top and bottom substrates, is theoretically well-described. For 1D LC polarization gratings working in transmission, the diffraction efficiency can be estimated theoretically based on the paraxial approximation for an infinite grating with $2\pi\lambda d / \left(n_0 \Lambda^2 \right) \ll 1$ [41,42]. For circularly polarized input light, a very high (theoretically 100%) efficiency into the first diffraction order is obtained when the half-wave condition is satisfied $\frac{\Delta n d}{\lambda} = 1/2$. Left- and right-handed circularly polarized light is diffracted in opposite directions (positive and negative diffraction order) and linearly polarized light is equally diffracted in both directions. The polarization state is conserved in the zeroth order while the handedness of the incident circular polarization is inversed for diffraction into the first order. This also holds when the half-wave condition is not satisfied.

At sufficiently high voltages, the 2D gratings studied here can be seen as a combination of two rotated 1D gratings that are decoupled by a layer of z-oriented LC in the bulk. The region with vertical director orientation in the middle of the cell (at high voltages) can approximately describe the behavior based on the two gratings associated with the top and bottom substrate respectively. Therefore at high voltages, the theoretical description presented above (for 1D gratings) can be used to understand the diffraction in the first diffraction orders (±1, 0) and (0, ±1), associated with the top and bottom substrate, and in the zero order (0, 0). The amount of light that is diffracted into the different orders depends on the voltage via the retardation. For vanishing retardation (at high voltages) light is concentrated in the zero order, with a polarization that is conserved. For circularly polarized incident light, light diffracted in the first diffraction orders is circularly polarized with a handedness that is inverted. These diffraction spots, associated with first order diffraction at the top and bottom grating respectively, are observed in position (-1, 0) and (0, -1) respectively. As can be seen in Figure 4a, a small amount of light is also diffracted towards (0, +1). This side effect might be related to small changes in the light polarization upon propagation through the LC layer, before reaching the grating associated with the bottom substrate. Diffraction into the (-1, +1) order, as can be seen in Figure 4 (at higher voltages), results from a combined effect from the grating at the top and bottom substrate: since the handedness is inversed by the diffraction towards the -1st order associated with the top substrate, the light can be diffracted a second time towards the +1st order associated with the bottom substrate. Thanks to this combined effect, the light diffracted into the (-1, +1) order has again the same polarization as the incident light. This diffraction spot is oriented along the k_y-axis (Figure 4).

The other diffraction spot ($-1/2$, +1/2) that appears along the k_y-axis, is especially important at lower voltages. The ($-1/2$, +1/2) diffraction order is dominant at 0 V and results from the Λ_y periodicity along the y-axis that is present at 0 V (Figures 2 and 3). As can be observed in the POM images in Figure 3, a structure with period Λ_y along the y-axis is formed at 0 V while the observed period at high voltages is $\Lambda_y/2$. This intuitively explains the presence of a strong diffraction order ($-1/2$, +1/2) along the k_y-axis at low voltages. To quantitatively describe the distribution of light into the different orders, detailed numerical simulations are necessary.

6. Conclusions

The diffraction characteristics and the observed microscopy images can be well understood with the help of numerical simulations for the director configuration. The formation of a 2D LC superstructure without disclinations is driven by energy minimization. Twist conflicts in the confining substrates are resolved with the help of out-of-plane director reorientation in the bulk of the layer and this concept is generally applicable for different rotation angles ψ between the alignment patterns at both substrates. Alternating line regions with (anti-)parallel top and bottom alignment lead to bulk regions with respectively close-to-vertical and roughly planar director orientation. Rotation angles $\psi = 60°$ and $120°$ give rise to hexagonal diffraction patterns that can be reversibly switched with the help of small electric fields.

Author Contributions: Conceptualization, I.N. and K.N.; Methodology, I.N. and B.B.; Software, I.N.; Validation, I.N. and K.N.; Formal analysis, I.N.; Investigation, I.N.; Data curation, I.N.; Writing—original draft preparation, I.N.; Writing—review and editing, K.N., and B.B.; Visualization, I.N.; Supervision, K.N. All authors have read and agreed to the published version of the manuscript.

Funding: This research was funded by the Flemish fund for scientific research (FWO Vlaanderen), (grant no. FWOOPR2021008601).

Data Availability Statement: Data may be provided by the authors upon request.

Acknowledgments: The authors want to acknowledge R. Declercq, J. Dewyspelaere, B. De Bruyn, and S. Goegebeur for their work on this topic during their bachelor project in 2021.

Conflicts of Interest: The authors declare no conflict of interest. The funders had no role in the design of the study; in the collection, analyses, or interpretation of data; in the writing of the manuscript, or in the decision to publish the results.

References

1. Woltman, S.J.; Jay, G.D.; Crawford, G.P. Liquid-crystal materials find a new order in biomedical applications. *Nat. Mater.* **2007**, *6*, 929–938. [CrossRef] [PubMed]
2. Miniewicz, A.; Gniewek, A.; Parka, J. Liquid Crystals for photonic applications. *Opt. Mater.* **2003**, *21*, 605–610. [CrossRef]
3. Lagerwall, J.P.F.; Scalia, G. A new era for liquid crystal research: Applications of liquid crystals in soft matter nano-, bio- and microtechnology. *Curr. Appl. Phys.* **2012**, *12*, 1387–1412. [CrossRef]
4. Schadt, M. Liquid crystal displays, LC-materials and LPP photo-alignment. *Mol. Cryst. Liq. Cryst.* **2017**, *647*, 253–268. [CrossRef]
5. Yaroshchuk, O.; Reznikov, Y. Photoalignment of liquid crystals: Basics and current trends. *J. Mater. Chem.* **2012**, *22*, 286. [CrossRef]
6. Chigrinov, V. Photoaligning and Photopatterning—A New Challenge in Liquid Crystal Photonics. *Crystals* **2013**, *3*, 149–162. [CrossRef]
7. Nys, I. Patterned surface alignment to create complex three-dimensional nematic and chiral nematic liquid crystal structures. *Liq. Cryst. Today* **2020**, *29*, 65–83. [CrossRef]
8. Jiang, M.; Yu, H.; Feng, X.; Guo, Y.; Chaganava, I.; Turiv, T.; Lavrentovich, O.D.; Wei, Q.-H. Liquid Crystal Pancharatnam—Berry Micro-Optical Elements for Laser Beam Shaping. *Adv. Opt. Mater.* **2018**, *6*, 1800961. [CrossRef]
9. Chen, P.; Lu, Y.Q.; Hu, W. Beam shaping via photopatterned liquid crystals. *Liq. Cryst.* **2016**, *43*, 2051–2061. [CrossRef]
10. Larocque, H.; Gagnon-Bischoff, J.; Bouchard, F.; Fickler, R.; Upham, J.; Boyd, R.W.; Karimi, E. Arbitrary optical wavefront shaping via spin- to-orbit coupling. *J. Optics* **2016**, *18*, 124002. [CrossRef]
11. Xiang, X.; Kim, J.; Escuti, M.J. Bragg polarization gratings for wide angular bandwidth and high efficiency at steep deflection angles. *Sci. Rep.* **2018**, *8*, 10–15. [CrossRef] [PubMed]
12. Chen, P.; Wei, B.Y.; Hu, W.; Lu, Y.Q. Liquid-Crystal-Mediated Geometric Phase: From Transmissive to Broadband Reflective Planar Optics. *Adv. Mater.* **2019**, *32*, 1903665. [CrossRef]
13. Kobashi, J.; Yoshida, H.; Ozaki, M. Planar optics with patterned chiral liquid crystals. *Nat. Photonics* **2016**, *10*, 389–392. [CrossRef]
14. Nys, I.; Stebryte, M.; Ussembayev, Y.Y.; Beeckman, J.; Neyts, K. Tilted chiral liquid crystal gratings for efficient large-angle diffraction. *Adv. Opt. Mater.* **2019**, *7*, 1901364. [CrossRef]
15. Hess, A.J.; Poy, G.; Tai, J.S.B.; Žumer, S.; Smalyukh, I.I. Control of light by topological solitons in soft chiral birefringent media. *Phys. Rev. X* **2020**, *10*, 031042. [CrossRef]
16. Nys, I.; Berteloot, B.; Poy, G. Surface stabilized topological solitons in nematic liquid crystals. *Crystals* **2020**, *10*, 840. [CrossRef]

17. Nys, I.; Beeckman, J.; Neyts, K. Fringe-field-induced out-of-plane reorientation in vertically aligned nematic spatial light modulators and its effect on light diffraction. *Liq. Cryst.* **2021**, *48*, 1516–1524. [CrossRef]
18. Guo, Y.; Jian, M.; Peng, C.; Sun, K.; Yaroshchuk, O.; Lavrentovich, O.; Wei, Q.-H. High-Resolution and High-Throughput Plasmonic Photopatterning of Complex Molecular Orientations in Liquid Crystals. *Adv. Mater.* **2016**, *28*, 2353–2358. [CrossRef]
19. Miskiewicz, M.N.; Escuti, M.J. Direct-writing of complex liquid crystal patterns. *Opt. Express* **2014**, *22*, 12691. [CrossRef]
20. Berteloot, B.; Nys, I.; Poy, G.; Beeckman, J.; Neyts, K. Ring-shaped liquid crystal structures through patterned planar photo-alignment. *Soft Matter* **2020**, *16*, 4999–5008. [CrossRef] [PubMed]
21. De Sio, L.; Roberts, D.E.; Liao, Z.; Nersisyan, S.; Uskova, O.; Wickboldt, L.; Tabiryan, N.; Steeves, D.M.; Kimball, B.R. Digital polarization holography advancing geometrical phase optics. *Opt. Express* **2016**, *24*, 18297–18306. [CrossRef]
22. Crawford, G.P.; Eakin, J.N.; Radcliffe, M.D.; Callan-Jones, A.; Pelcovits, R.A. Liquid-crystal diffraction gratings using polarization holography alignment techniques. *J. Appl. Phys.* **2005**, *98*, 123102. [CrossRef]
23. Provenzano, C.; Pagliusi, P.; Cipparrone, G. Electrically tunable two-dimensional liquid crystals gratings induced by polarization holography. *Opt. Express* **2007**, *15*, 5872–5878. [CrossRef] [PubMed]
24. Wang, M.; Li, Y.; Yokoyama, H. Artificial web of disclination lines in nematic liquid crystals. *Nat. Commun.* **2017**, *8*, 389–392. [CrossRef] [PubMed]
25. Nys, I.; Beeckman, J.; Neyts, K. Switchable 3D liquid crystal grating generated by periodic photo-alignment on both substrates. *Soft Matter* **2015**, *11*, 7802–7808. [CrossRef]
26. Nys, I.; Nersesyan, V.; Beeckman, J.; Neyts, K. Complex liquid crystal superstructures induced by periodic photo-alignment at top and bottom substrate. *Soft Matter* **2018**, *14*, 6892–6902. [CrossRef]
27. Nersesyan, V.; Nys, I.; Van Acker, F.; Wang, C.-T.; Beeckman, J.; Neyts, K. Observation of symmetry breaking in photoalignment-induced periodic 3D LC structures. *J. Mol. Liq.* **2020**, *306*, 112864. [CrossRef]
28. Honma, M.; Toyoshima, W.; Nose, T. Bistable liquid crystal device fabricated via microscale liquid crystal alignment. *J. Appl. Phys.* **2016**, *120*, 143105. [CrossRef]
29. Honma, M.; Nose, T. Twisted nematic liquid crystal polarization grating with the handedness conservation of a circularly polarized state. *Opt. Express* **2012**, *20*, 18449–18458. [CrossRef] [PubMed]
30. James, R.; Willman, E.; Fernández, F.A.; Day, S.E. Finite-Element Modeling of Liquid-Crystal Hydrodynamics with a Variable Degree of Order. *IEEE Trans. Electron Devices* **2006**, *53*, 1575–1582. [CrossRef]
31. James, R.; Willman, E.; Ghannam, R.; Beeckman, J.; Fernández, F.A. Hydrodynamics of fringing-field induced defects in nematic liquid crystals. *J. Appl. Phys.* **2021**, *130*, 134701. [CrossRef]
32. De Gennes, P.G.; Prost, J. *The Physics of Liquid Crystals*, 2nd ed.; Oxford University Press: Oxford, UK, 1993; pp. 76–78.
33. Landau, L.D.; Lifshitz, E.M.; Pitaevskii, L.P. *Statistical Physics*, 3rd ed.; Pergamon Press: Oxford, UK, 1980; pp. 440–442.
34. Coles, H.J. Laser and Electric Field Induced Birefringence Studies on the Cyanobiphenyl Homologues. *Mol. Cryst. Liq. Crys.* **1978**, *49*, 67. [CrossRef]
35. Parry-Jones, L.A.; Elston, S.J. Flexoelectric switching in a zenithally bistable nematic device. *J. Appl. Phys.* **2005**, *97*, 093515. [CrossRef]
36. Poy, G.; Zumer, S. Physics-based multistep beam propagation in inhomogeneous birefringent media. *Opt. Express* **2020**, *28*, 24327–24341. [CrossRef] [PubMed]
37. Pancharatnam, S. Generalized theory of interference, and its applications. *Proc. Indian Acad. Sci. A* **1956**, *44*, 247–262. [CrossRef]
38. Berry, M.V. The adiabatic phase and Pancharatnam's phase for polarized light. *J. Mod. Opt.* **1987**, *34*, 1401–1407. [CrossRef]
39. Varanytsia, A.; Posnjak, G.; Mur, U.; Joshi, V.; Darrah, K.; Musevic, I.; Copar, S.; Shien, L.C. Topology-commanded optical properties of bistable electric- field-induced torons in cholesteric bubble domains. *Scien. Rep.* **2017**, *7*, 16149. [CrossRef]
40. Kim, J.H.; Huh, J.W.; Oh, S.W.; Ji, S.M.; Jo, Y.S.; Yu, B.H.; Yoon, T.H. Bistable switching between homeotropic and focal-conic states in an ion-doped chiral nematic liquid crystal cell. *Opt. Express* **2017**, *25*, 29180–29188. [CrossRef]
41. Tervo, J.; Turunen, J. Paraxial-domain diffractive elements with 100% efficiency based on polarization gratings. *Opt. Lett.* **2000**, *25*, 785–786. [CrossRef]
42. Escuti, M.J.; Jones, W.M. Polarization-Independent Switching with High Contrast from A Liquid Crystal Polarization Grating. *SID Symp. Digest.* **2006**, *37*, 1443–1446. [CrossRef]

Article

Polymer Stabilization of Uniform Lying Helix Texture in a Bimesogen-Doped Cholesteric Liquid Crystal for Frequency-Modulated Electro-Optic Responses

Chia-Hua Yu [1], Po-Chang Wu [2] and Wei Lee [2,*]

[1] College of Photonics, National Yang Ming Chiao Tung University, Guiren Dist., Tainan 711010, Taiwan; chyu1029@gmail.com

[2] Institute of Imaging and Biomedical Photonics, College of Photonics, National Yang Ming Chiao Tung University, Guiren Dist., Tainan 711010, Taiwan; jackywu@nycu.edu.tw

* Correspondence: wei.lee@nycu.edu.tw; Tel.: +886-6-303-2121 (ext. 57826); Fax: +886-6-303-2535

Abstract: A polymer network (PN) can sustain the uniform lying helix (ULH) texture in a binary cholesteric liquid crystal (LC) comprising a calamitic LC and a bimesogenic LC dimer. Upon copolymerization of a bifunctional monomer with a trifunctional monomer at a concentration of 5 wt% to create the desired polymer network structure, the PN-ULH was obtained with high stability and recoverability even when cycles of helical unwinding-to-rewinding processes were induced after the electrical or thermal treatment. Utilizing dielectric spectroscopy, the flexoelectric-polarization-dominated dielectric relaxation in the PN-ULH state was characterized to determine two frequency regions, $f < f_{\text{flexo}}$ and $f > f_{\text{di}}$, with pronounced and suppressed flexoelectric effect, respectively. It is demonstrated that the cell in the PN-ULH state can operate in the light-intensity modulation mode by the flexoelectric and dielectric effects at $f < f_{\text{flexo}}$ and phase-shift mode by the dielectric effect at $f > f_{\text{di}}$. Moreover, varying the voltage frequency from $f < f_{\text{flexo}}$ to $f > f_{\text{di}}$ results in a frequency dispersion of transmittance analogous to that of flexoelectric-polarization-dominated dielectric relaxation. The unique combination of the bimesogen-doped cholesteric LC with a stable and recoverable PN-ULH texture is thus promising for developing a frequency-modulated electro-optic device.

Keywords: cholesteric liquid crystal; uniform lying helix; polymer network; frequency modulation; electro-optic response; mesogenic dimer; flexoelectric effect; dielectric effect

Citation: Yu, C.-H.; Wu, P.-C.; Lee, W. Polymer Stabilization of Uniform Lying Helix Texture in a Bimesogen-Doped Cholesteric Liquid Crystal for Frequency-Modulated Electro-Optic Responses. *Materials* **2022**, *15*, 771. https://doi.org/10.3390/ma15030771

Academic Editor: Ingo Dierking

Received: 16 December 2021
Accepted: 18 January 2022
Published: 20 January 2022

Publisher's Note: MDPI stays neutral with regard to jurisdictional claims in published maps and institutional affiliations.

1. Introduction

Uniform lying helix (ULH) is a kind of cholesteric liquid crystal (CLC) texture whose helical pitch is shorter than the wavelengths of visible light and is uniformly aligned along a preferred direction in the substrate plane of confining geometry. When a voltage is applied across the ULH helix, two types of helix reorientation can be generated by voltage-induced flexoelectric and dielectric effects. At a moderate voltage where the dielectric effect is negligible, the flexoelectric coupling of LC molecules with the electric field produces a periodic splay-bend deformation, giving rise to the in-plane deviation of the ULH optic axis. The magnitude of such a deviation angle induced by the flexoelectric effect is in principle a linear function of the voltage amplitude and is also in positive correlation with the average value of the splay and bend flexoelectric coefficients [1]. If the voltage is beyond a critical threshold such that the dielectric effect is electrically induced, the helix of the CLC with positive dielectric anisotropy is gradually unwound with ascending voltage amplitude, and the LC configuration is eventually sustained in the helix-free homeotropic state at high voltage, leading to the change in effective birefringence from $(n_e + n_o)/2$ in the field-off ULH state to n_o in the voltage-sustained homeotropic state, where n_e and n_o are the extraordinary and ordinary refractive indices of the LC, respectively [2]. Owing to the uniqueness, a CLC in the ULH state can be regarded as an electrically tunable birefringent medium with the optic axis along the helical axis in the field-off state for modulating the intensity and phase retardation of polarized light under a pair of crossed polarizers. Since the chiral-flexoelectric effect with sub-millisecond response time was first exploited by Patel and Meyer in 1987 [3], short-pitch CLCs with ULH alignment have drawn a great deal of interest as promising fast-response optoelectronic materials for high-definition displays with a field-sequential color driving scheme, virtual/augmented reality head-mounted elements, and high-speed spatial light modulators in glasses-free 3D displays, as well as light detection and ranging (LiDAR) sensors [4–7]. Potential applications of ULH with positive dielectric anisotropy based on the voltage-induced dielectric effect have also been suggested as electrically tunable phase retarders [8] and lasers [9].

From the point of view of device fabrication, the formation of a stable and defect-free ULH configuration in a sandwich-type cell is problematic, in that a CLC enclosed in such a simple confining geometry favors stabilization in the Grandjean planar state or the focal conic state to minimize the free energy. The most widely used aligning technique is to apply AC voltage across a cell upon cooling through the isotropic-to-CLC phase transition [3]. The uniformity of ULH formed with this electric field approach has been improved by mechanically shearing the cell substrates simultaneously [10,11] or by modifying the surface alignment conditions with alternate stacking of planar and homeotropic alignment [12], or with helical-pitch-matched cholesteric alignment layer [13]. Other electric field approaches—including those based on voltage-induced electrohydrodynamic instability [14,15], flexoelectric effect [16], textural transition [17], electro-thermal effect [18], and adoption of tri-electrode configurations [19,20]—have been successively proposed for generating ULH alignment in a planar-aligned cell. Nevertheless, most ULH textures as prepared by the above-mentioned voltage pretreatments would be irreversibly destroyed or even transferred back to the most stable Grandjean planar texture after the CLC helix is partially or completely unwound by external voltages or the phase transition often occurs unwantedly by thermal variation; hence, making them incompatible for utilization of the flexoelectric or dielectric switching in extensive electro-optic applications. As such, attention has been paid to adding photocurable precursors into CLCs for forming polymer networks via photopolymerization to effectively stabilize the ULH structure for the flexoelectro-optic effect [2,21–25].

In a practical situation, the flexoelectric coefficients of conventional rod-like LCs are relatively small, so the extent of flexoelectric switching would be insufficient, and the electro-optic characteristics are dominated primarily by the dielectric effect. To promote flexoelectro-optic performance, specific bent-core and bimesogenic LCs with inherently high flexoelectric coefficients and low dielectric anisotropy have been synthesized to lower the operating voltage and enlarge the voltage-driven deviation angle of the optic axis [11,21,26,27]. While these bimesogenic compounds are not as ubiquitous or universal as rod-like LCs and their LC mesophases exist mostly at high temperatures, pronounced flexoelectric effect has alternatively been revealed at room temperature in a binary CLC mixture consisting of bent-core bimesogens incorporated into a commercial rod-like CLC with low dielectric anisotropy [16,22,28]. As the material properties required are a trade-off, it seems difficult to embrace comparably strong flexoelectric and dielectric effects in a CLC ULH cell.

The bent-core LC dimer CB7CB, offering alluring material properties of high flexoelectric coefficients and low bend elastic constant, is the first bimesogneic compound discovered to exhibit the twist-bend nematic phase [29]. More attractively is the fact that mixing CB7CB with a given rod-like CLC has been proven unique, allowing the resultant binary CLC mixture to exhibit the desired flexoelastic ratio in a wide temperature range. Superior features based on CB7CB-doped CLCs have been suggested for a diverse range of photonic and electro-optic applications, especially fast and pronounced flexoelectro-optic responses in the ULH state of CB7CB-doped short-pitch CLCs with promoted flexoelectric coefficients [21,22] and the electrically tunable reflection bandgap in a wide range (from UV to visible and infrared) in the heliconical state of a CB7CB-doped long-pitch CLC whose elastic ratio of bend to elastic constants smaller than 1 [30]. Recently, we have developed a binary CLC mixture containing a bimesogenic LC dimer (CB7CB) with large flexoelectric coefficients, a conventional rod-like nematic LC (E7) with large dielectric anisotropy, and a chiral dopant. By means of dielectric spectroscopy and transmission spectroscopy, as well as textural observations, we have established an electrical approach to generating a stable ULH texture and clarified the unusual frequency-modulated textural switching mechanism in terms of the frequency-*dependent* flexoelectric effect and frequency-*independent* dielectric effect. In the present work, we refer to the alluring material feature specific to the CB7CB-doped E7 CLC and aimed to develop a polymer-network (PN)-stabilized ULH texture for implementing frequency-modulated electro-optic responses via the dielectric and flexoelectric effects. Unlike previously developed polymer networks employing a photocurable monomer with particular ultraviolet (UV) exposure conditions or with high concentrations [2,21–25], the proposed one in this study was constructed by a monomer mixture containing a bifunctional monomer RM257 and a trifunctional monomer TMPTA, permitting the ULH texture to be highly stable and reversibly recoverable at a low concentration of 5-wt% in a total. So far, the polymer system involving the copolymerization of RM257 and TMPTA has widely been applied to the stabilization of blue phases in wide temperature ranges [31,32], but it has not been adopted for stabilizing the ULH texture until now. To demonstrate the unique combination of RM257 and TMPTA, another PN-ULH texture with RM257 alone was fabricated as a comparative counterpart. The stability in the static state and recoverability after stimuli-driven textural or phase transition of the two PN-ULH textures with distinct polymer structures were discussed by textural observation along with transmission and dielectric measurements. Furthermore, the frequency dependence of the flexoelectric and dielectric strengths was manifested by the real-part dielectric spectra. Voltage and frequency-dependent transmission spectra were revealed to clarify the electro-optic responses of the PN-ULH in specific frequency ranges and to demonstrate the feasibility for complying with light intensity and phase modulations by the voltage amplitude and frequency.

2. Materials and Methods

2.1. Materials

Materials used in this study include the calamitic nematic LC E7, bent-core LC dimer CB7CB ($1'',7''$-bis(4-cyanobiphenyl-$4'$-yl)heptane), chiral additive R5011, photocurable monomers RM257 (1,4-Bis-[4-(3-acryloyloxypropyloxy)benzoyloxy]-2-methyl-benzene) and TMPTA (1,1,1-trimethylolpropane triacrylate), and the photoinitiator Irg184 (1-hydroxy cyclohexane benzophenone). E7 was provided by Daily Polymer Corp., Kaohsiung, Taiwan, and TMPTA was purchased from Alfa Aesar, Ward Hill, MA, USA. The others were obtained from Jiangsu Hecheng Display Technology Corp., Nanjing, China. All the above-mentioned materials with chemical structures, as illustrated in Figure 1, were used as received without further purification. E7 with a rod-like molecular shape is a cyano-based and four-component eutectic mixture with a clearing temperature $T_c = 59\,^\circ\mathrm{C}$, birefringence $\Delta n = 0.225$ (measured at the wavelength $\lambda = 589$ nm and temperature $T = 20\,^\circ\mathrm{C}$), and positive dielectric anisotropy of $\Delta\varepsilon = +14.3$ (measured at the frequency $f = 1$ kHz and $T = 20\,^\circ\mathrm{C}$), as given in the datasheet. CB7CB is a bent mesogenic dimer whose material properties have been found particularly to exist as a nematic twist-bend phase (N_{TB}) at $T < 99\,^\circ\mathrm{C}$ [33], showing a high flexoelectric coefficient of ~31 pC/m and low dielectric anisotropy around +1–2 [21]. R5011 exhibits right-handed chirality with a relatively high helical twisting power (HTP) over 100 μm^{-1}. RM257 is a bifunctional liquid crystalline monomer with a rod-like molecular shape, and TMPTA is a trifunctional monomer with three acrylate ester groups. The function of the photoinitiator Irg184, capable of generating free radicals under UV exposure, is to accelerate the photopolymerization process of monomers.

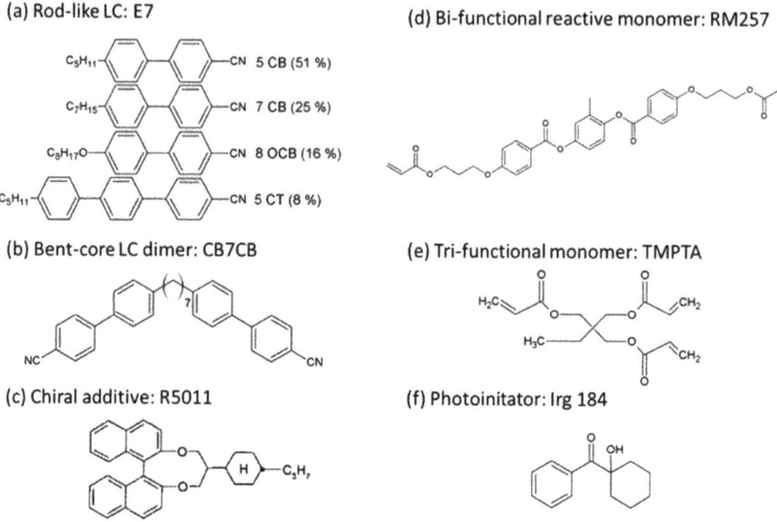

Figure 1. Chemical structures of (**a**) E7, (**b**) CB7CB, (**c**) R5011, (**d**) RM257, (**e**) TMPTA, and (**f**) Irg 184.

2.2. Sample Preparations

A binary achiral LC mixture, consisting of 55-wt% E7 and 45-wt% CB7CB, was blended with R5011 at a weight percentage of c_{R5011} = 3.8-wt% to obtain a short-pitch CLC with a reflective bandgap lying in the UV light regime in the Grandjean planar state and with bimesogen-enhanced flexoelectro-optic responses in the ULH state. The HTP of R5011 in the 45-wt% CB7CB-containing E7 was proven to be 145 μm^{-1} [16] so that the helical pitch length of the resultant CLC is 181 nm as calculated by $(HTP \times c_{R5011})^{-1}$. Two monomer/CLC precursors, designated RM50 and RM25TM25, were prepared by incorporating the CLC with 5.0-wt% RM257 and 0.2-wt% Irg 184, and with 2.5-wt% RM257, 2.5-wt% TMPTA, and 0.2-wt% Irg 184, respectively. After thorough stirring for 2 h at 100 °C to allow homogeneous blending. Each monomer/CLC precursor in the isotropic phase at 100 °C was then injected via the capillary action into a planar-aligned cell (Chiptek, Miaoli, Taiwan) with a cell gap of 5.0 ± 0.5 μm and overlapped electrode area of 0.25 cm^2. The inner surfaces of the two substrates were spin-coated with the polyimide DL-2360 as the planar aligning agent and rubbed in anti-parallel directions to impose the planar surface anchoring of LC molecules with a pretilt angle of 4–6°. The phase sequences of RM25TM25 and RM50 precursors as examined by optical textures in the cooling process from 70 °C to 0 °C are Isotropic—58 °C—blue phase—17 °C—CLC—7 °C—SmA* and Isotropic—61 °C—blue phase—18 °C—CLC—11 °C—SmA*, respectively. Note that the obtained blue phase within a temperature range over 40 °C is not permanently stable because it would transfer to the CLC phase after applying a significantly high voltage to unwind the helical structure or by heating from a temperature in the SmA* phase.

2.3. Formation of a Polymer-Network Uniform Lying Helix (PN-ULH) Texture

Figure 2 illustrates the procedure for forming a given polymer structure in an RM25TM25-CLC or an RM50-CLC cell with ULH alignment. All steps were performed at T = 25 °C stabilized with a temperature controller (Linkam Scientific, T95-PE, Surrey Tadworth, UK). First, a 90-V$_{rms}$ voltage at 5 kHz was applied across the cell to sustain the LC orientation in the homeotropic state (Figure 2a,d). Next, referring to the established electric field approach in our previous work [16], defect-free, monodomain ULH alignment was electrically induced via the flexoelectric effect by switching the frequency of 90-V$_{rms}$ voltage directly from 5 kHz to 100 Hz (Figure 2b,e). By decreasing the voltage slowly from 90 V$_{rms}$ to 0 V$_{rms}$ to avoid defect generation, the well-aligned ULH texture was stably preserved, and the cell at zero voltage was illuminated with UV light at $\lambda = 365$ nm and intensity of 6 mW·cm^{-2} derived from a Panasonic Aicure UJ35 LED spot-type UV curing system to allow chain polymerization of monomers. After 8 min of UV exposure, two PN-ULH textures were obtained with distinct polymer structures constructed by 2.5-wt% RM257 and 2.5-wt% TMPTA in the RM25TM25-CLC cell (Figure 2c) and by 5.0-wt% RM257 in the RM50-CLC cell (Figure 2f).

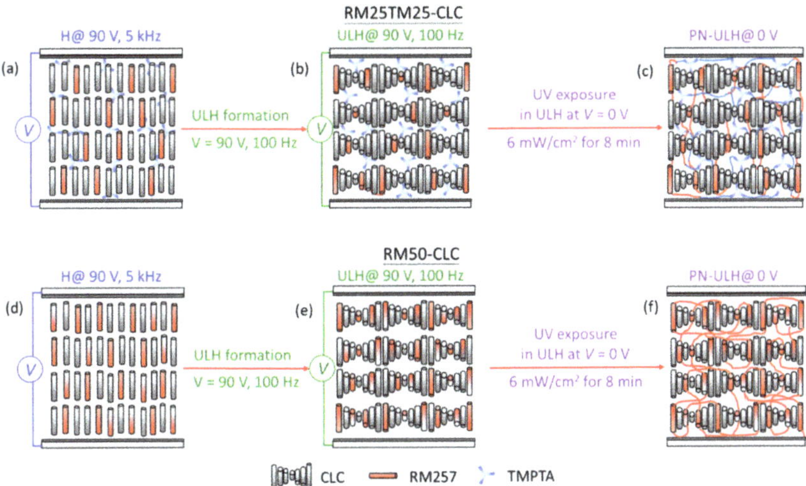

Figure 2. Schematic illustrations of steps for fabricating a PN-ULH texture in (**a–c**) an RM25TM25-CLC cell and (**d–f**) an RM50-CLC cell. (**a,d**) Electrically sustaining LC orientation by a 90-V$_{rms}$ voltage at 5 kHz; (**b,e**) forming ULH texture by switching the frequency of 90-V$_{rms}$ voltage from 5 kHz to 100 Hz; (**c,f**) constructing polymer network via photopolymerization after 8 min of 6 mW·cm^{-2} UV exposure.

2.4. Measurements

Measurements were performed at $T = 25\,°C$ unless specified. The types of mesophases and CLC textures were identified by optical images using a polarizing optical microscope (BX51-P, Olympus, Tokyo, Japan) in transmission mode with a 10× objective lens. Images were taken by a digital camera (Olympus XC30, Tokyo, Japan) with a resolution of 2080 × 1544 pixels mounted on the microscope. To evaluate optical and electro-optic characteristics of a cell, transmission spectra in the visible range of 400 nm–750 nm were acquired with a high-speed fiber-optic spectrometer (Ocean Optics HR2000+, Shanghai, China) in conjunction with a halogen light source (Ocean Optics HL2000, Shanghai, China). Time-dependent transmission features were monitored using a He–Ne laser operating at $\lambda = 632.8$ nm as the light source. All the experiments mentioned above were accomplished under crossed polarizers, and the angle between the helical axis of ULH in the field-off state and the transmission axis of either polarizer is defined as the polarizing angle α. An arbitrary function generator (Tektronix AFG-3022B, Beaverton, OR, USA) was connected to an amplifier (TREK Model 603, New Taipei, Taiwan) to permit electric field (in square waves) applied across a cell in a wide tunable range of 0 V–125 V. Dielectric spectra were obtained using a precision LCR meter (Agilent-E4980A, Santa Clara, CA, USA) with a measurable frequency range from 20 Hz–2 MHz. The probe voltage was as low as 0.5 V$_{rms}$ to avoid reorienting LC molecules in the dielectric measurement.

3. Results and Discussion

3.1. Stability and Recoverability of PN-ULH Textures

Figure 3 shows optical textures in the static state of the RM25TM25- and RM50-CLC cells as prepared and after treating with a cycle of voltage-induced helical unwinding. In the case of the RM25TM25-CLC (Figure 3a), the uniform dark image at $\alpha = 0°$ and bright image at $\alpha = 45°$, attributable to the birefringent effect, indicate that the PN-ULH texture thus-formed following the said procedure in Figure 2a–c was stable and of high quality with a unidirectional helical axis parallel to the rubbing direction. By applying a significantly high voltage at $V = 100$ V_{rms} and $f = 5$ kHz across the cell thickness, enabling the ULH helix to be unwound and LC orientation sustained in the homeotropic state, the dark (bright) image at $\alpha = 0°$ (45°) at $V = 0$ V was virtually unchanged after the voltage removal (Figure 3a), connoting that the PN-ULH was retained as it was prior to the voltage application. In comparison, the RM50-CLC as prepared revealed a stable PN-ULH texture as implied by the appearance of high dark-to-bright contrast between images at $\alpha = 0°$ and 45° in Figure 3b. For cells prepared because more stripe defects were generated during the generation of ULH in RM50-CLC step prior to photopolymerization, the optical texture of the resulting PN-ULH (Figure 3b) is somewhat non-uniform as compared with that of RM25TM25-CLC (Figure 3a). Nevertheless, when a cycle of helical unwinding and rewinding process was performed by the switching of a 5-kHz voltage from 100 V_{rms} to 0 V, the Grandjean planar texture instead of ULH was generated together with those already existing stripe defects at $V = 0$ V, leading to equally dark optical images with visible tiny-sized domains at $\alpha = 0°$ and 45° (Figure 3b). The validity of the Grandjean planar texture in the RM50-CLC sample after voltage treatment has been double-checked by the visible light transmission spectrum without any polarizer, which shows transmittance of ~80% on average compared to that of ITO glass used (data not shown). The resultant Grandjean planar texture with high transparency in visible light might be attributable to the CB7CB-doped CLC host with matched bend and twist elastic constants so that the nucleation process for the transition from the transient planar state to the initial Grandjean state was reduced [34]. Subsequently, a continuous voltage waveform comprising $V = 100$ V_{rms} at 5 kHz for 3 s and V = 0 for 7 s in a period was designated to repeatedly unwind and rewind the CLC helix. Figure 4 depicts the time-varying optical transmission ($\lambda = 632.8$ nm) of the CLC cells in response to the voltage wave train. The reliability and recoverability of the PN-ULH in the RM25TM25-CLC cell can be ascertained by the results in Figure 4a, showing equally high contrast between transmittances at ~1% in durations with $V = 100$ V_{rms} at 5 kHz (i.e., the V-sustained H state) and ~70% in those with $V = 0$ V (i.e., the stable ULH state) in every period. It is worth noting that the PN-ULH in the RM25TM25-CLC after cycles of voltage treatments can stably be maintained for at least 2 weeks. In contrast, Figure 4b indicates that driving the RM50-CLC with the designated voltage waveform resulted in nearly invariant transmittance at ~1% over time. Because the optic axes in the Grandjean planar and homeotropic states were perpendicular to the substrate plane and along the propagation direction of normal incidence of light, no phase retardation was accumulated, and the intensity of light passing through a CLC cell in either state under crossed polarizers principally vanished. Consequently, it is suggested that the helices in the RM50-CLC, firstly unwound by the 100-V_{rms} voltage, tended to rewind to form the Grandjean planar texture rather than the ULH after removing the voltage. Although the ULH configuration in the RM50-CLC can be electrically regenerated following the steps illustrated in Figure 2d,e, it was still unstable after a cycle of helical relaxation from unwinding to rewinding. The results described above support that the polymer framework constructed by the copolymerization of 2.5-wt% RM257 and 2.5-wt% TMPTA was significant, allowing the ULH texture to be optically stable in the static state, electrically switchable to the homeotropic state, and recoverable from the helical unwinding state. However, for the counterpart with the TMPTA content fully replaced by RM257, the polymer structure made from the bifunctional reactive monomer alone (at 5.0-wt%) in the RM50-CLC became insufficient to sustain the ULH helix, causing it to become unstable after the CLC helix was unwound. This underlines the importance of TMPTA, which has been found to play the role of promoting the crosslinking density and introducing rigidity to the resulting polymer network with RM257 [31,32].

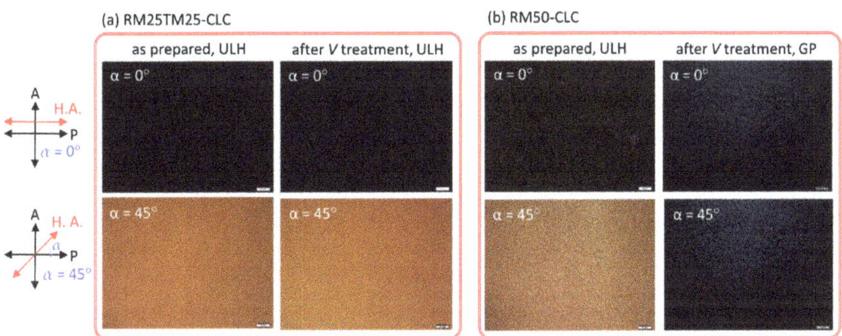

Figure 3. Optical images under crossed polarizers in the static state at $\alpha = 0°$ and 45° of (**a**) RM25TM25-CLC (**b**) RM50-CLC as prepared without any prior stimulus and after a voltage treatment at $T = 25\ °C$. The voltage treatment entails a 100-V_{rms} voltage at 5 kHz across a cell. Scale bar: 50 μm. P, A and H.A. represent the orientations of the transmission axis of the linear polarizer, the analyzer, and the ULH helical axis, respectively.

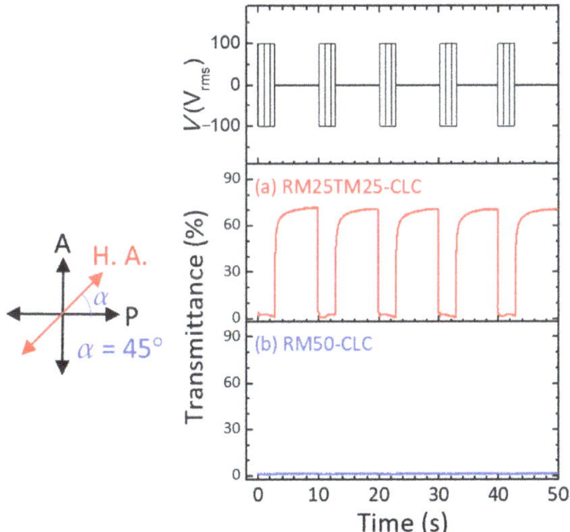

Figure 4. Dynamic transmission at $\lambda = 632.8$ nm of (**a**) RM25TM25 and (**b**) RM50 CLC cells (azimuthally oriented at 45° between crossed polarizers) subjected to voltage pulses consisting of $V = 100\ V_{rms}$ at 5 kHz for 3 s and $V = 0$ for 7 s successively in a pulse.

Furthermore, the thermal stability of PN-ULH textures and their recoverability after thermally induced phase transition were assessed in accordance with simultaneous measurements of optical images and real-part dielectric permittivity (ε') at varying temperatures as shown in Figure 5. Here, the ULH texture in the RM50-CLC was regenerated following the electric field approach illustrated in Figure 2d,e. The two cells initially with PN-ULH textures were first heated progressively from 25 °C to 70 °C at a rate of 1 °C/min. As they were unraveled by the bright appearances in the optical images at $\alpha = 45°$ (e.g., Figure 5a at 25 °C), the PN-ULH textures in this heating process were preserved until $T > T_{iso} = 60\ °C$, where the CLCs were in the isotropic phase with dark appearances in optical images (data not shown). The value of ε' in the partially ordered CLC ULH state, equaling to $(\varepsilon_{||} + \varepsilon_{\perp})/2$, is greater than $(\varepsilon_{||} + 2\varepsilon_{\perp})/3$ in the disordered isotropic phase, where $\varepsilon_{||}$ and $\varepsilon_{\perp}$ are respectively the parallel and perpendicular components of dielectric permittivity of LC molecules.

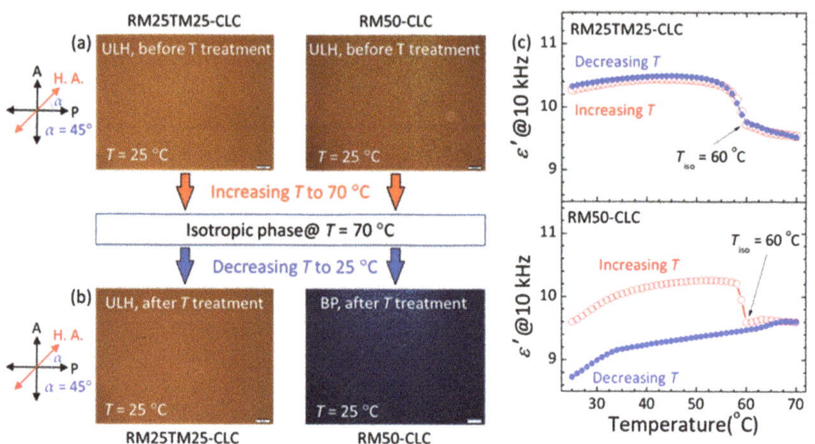

Figure 5. Optical images under crossed polarizers at $\alpha = 45°$ and $T = 25\ °C$ of RM25TM25-CLC RM50-CLC cells (**a**) before and (**b**) after a cycle of thermal treatment. (**c**) Temperature-dependent real-part dielectric permittivity at $f = 10\ kHz$ of the two cells, as measured by first increasing T from 25 °C to 70 °C, followed by decreasing T back to 25 °C with a heating/cooling rate of 1 °C/min. The subscript "iso" represents "isotropic." Scale bar: 50 μm.

The CLC-to-isotropic phase transition behavior upon heating can be depicted by the decreasing ε' with elevating T, particularly in the temperature range of 56–60 °C (red open circles in Figure 5c), and the exact value of $T_{iso} = 60\ °C$ be quantified by the first-order derivative of T-dependent ε'. When the cells were subsequently cooled down from 70 °C to 25 °C, the PN-ULH configuration in the RM25TM25-CLC cell was recovered with an identical bright texture, but in RM50-CLC it disappeared and, instead, the blue phase with a dark-blue texture emerged after the transition from the isotropic phase at $T < T_{iso} = 60\ °C$ (Figure 5b). Therefore, the T-dependent ε' curves, measured in the heating and cooling processes, overlapped in the RM25TM25-CLC but behaved differently in the RM50-CLC (Figure 5c). It is worth mentioning that the PN-ULH in the RM25TM25 cell could also be retained after heating from the SmA* to CLC phase. While for the blue phase in the RM50 cell, it was still unstable and became a Grandjean planar CLC by electrically unwinding the helical structure or by thermally induced phase transition from the SmA* phase.

To inspect morphologies of the polymer structures in the two CLCs, each cell was opened, and the two substrates were rinsed with acetone to remove LCs and then subjected to a scanning electron microscope (SEM; Hitachi S-4700 Type II, Tokyo, Japan). Figure 6 shows SEM images from the top-view of one of the two substrates of each PN-ULH CLC cell. One can clearly identify that a periodical polymer network was obtained on the surface of the RM25TM25 cell with a regular periodicity of ~96 nm equal to half of the helical pitch of the CLC host used (Figure 6a), but no such a periodical morphology was found at the substrate surface of the RM50-CLC (Figure 6b). The same result regarding the existence of a surface-localized polymer network was observed for the other substrate of a cell. With fixed UV exposure conditions (i.e., 6 mW·cm^{-2} for 8 min) in this work, the polymer structure stemming from 5.0-wt% RM257 was primarily formed in the main, but from 2.5-wt% RM257 and 2.5-wt% TMPTA was distributed not only in volume but at the surface. So far, depending on the constituents of the CLC hosts and the UV exposure conditions, a good number of studies have demonstrated stable PN-ULH textures enabled by polymer networks locally at one or two substrate surfaces [21,22], uniformly in volume, non-uniformly in volume, and at surfaces [2,23,24], or even with polymer walls [25] by using a kind of reactive monomer, such as RM257 or BAB6, in a wide span of concentrations between 3.5-wt% and 25-wt%. It is likely that for the binary CB7CB/E7 CLC host used in this study, a stable and recoverable PN-ULH texture could be created by increasing the concentration of RM257 over 5.0-wt%, but the trade-off would be the degraded electro-optic performance, especially at the expense of the increased operation voltage and residual birefringence. As a result, the RM25TM25-CLC with a stable and recoverable PN-ULH texture in a wide temperature range was adopted to explore electro-optic responses based on the flexoelectric and dielectric effects in the next section.

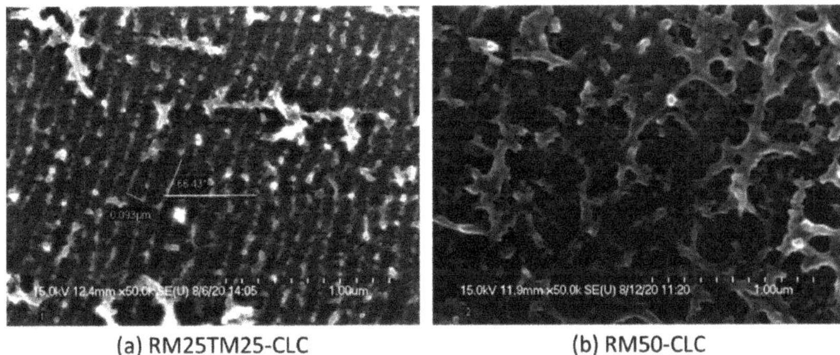

(a) RM25TM25-CLC   (b) RM50-CLC

Figure 6. SEM images from a top-view of polymer network morphologies on a substrate surface of (**a**) RM25TM25 and (**b**) RM50 CLCs with PS-ULH textures.

3.2. Frequency-Modulated Electro-Optic Responses

Prior to electro-optic investigations, we first discussed the extent of the flexoelectric effect, modified by the bimesogen CB7CB in the RM25TM25 CLC, via the real-part dielectric spectrum ($\varepsilon'(f)$) of the cell in the PN-ULH state according to the flexoelectric contribution to the dielectric permittivity. For a CLC in the ULH state with a strong flexoelectric effect, two types of dielectric relaxation behavior by the coupling of molecules to the electric field can be induced in distinct frequency regimes. The primary is connected to the orientational polarization by the molecular rotation around the short axis, which typically occurs beyond MHz in conventional thermotropic LCs and thus is excluded in this study. The secondary from chiral-flexoelectric polarization with a relaxation frequency between several hundred hertz and a few kilohertz is attributable to the splay-bend distortion through flexoelectric coupling to the electric field [28]. As plotted in Figure 7, the resolved frequency dispersion of dielectric permittivity with a relaxation frequency at ~710 Hz in the frequency range of 20 Hz–30 kHz from the experimental data (open circles) is undoubtedly attributable to the flexoelectric polarization, which has been theoretically explained in [28] and experimentally demonstrated in [16]. Such a Debye-like dielectric relaxation behavior contributed by flexoelectric polarization can be expressed in terms of the relaxation frequency (f_R), dielectric permittivities at low- (ε_L) and high-frequency (ε_H) limits as:

$$\varepsilon\prime(f) = \varepsilon_H + \frac{\varepsilon_L - \varepsilon_H}{1 + (f/f_R)^2} \tag{1}$$

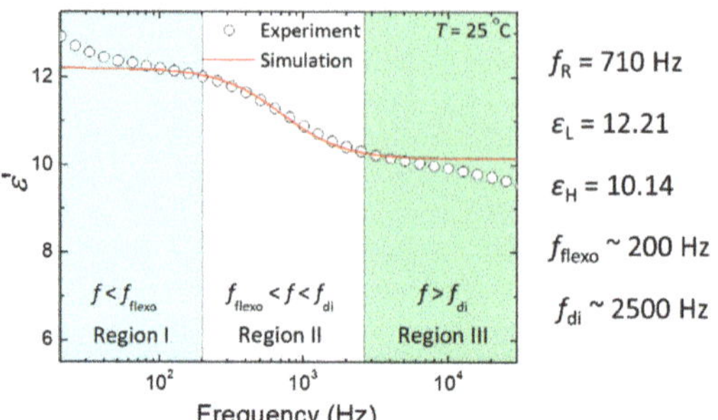

Figure 7. Real-part dielectric spectra $\varepsilon'(f)$ of the RM25TM25-CLC cell in the PN-ULH state at $T = 25\,^{\circ}\mathrm{C}$. The simulated $\varepsilon'(f)$ curve (solid red line) is plotted with fitting parameters of $f_R = 710$ Hz, $\varepsilon_L = 12.21$ and $\varepsilon_H = 10.14$, as deduced by fitting the experimental data (open circles) into Equation (1).

By fitting the experimental data of Figure 7 to Equation (1), we obtained $f_R = 710$ Hz, $\varepsilon_L = 12.21$ and $\varepsilon_H = 10.14$. The dielectric strength (ε_{flexo}) of this relaxation proportional to e^2/K [28] is equal to 2.07 as calculated by $\varepsilon_{flexo} = \varepsilon_L - \varepsilon_H$, where e and K are average values of splay and bend flexoelectric coefficients and elastic constants, respectively. Using these fitting parameters and referring to Equation (1), the original profile of the dielectric relaxation was simulated and plotted in Figure 7 (solid red line), which agrees with the experimental data. Note that the slight increase (decrease) of dielectric permittivity with decreasing (increasing) frequency at $f < 100$ Hz ($f > 4$ kHz) in the experimental dielectric spectrum is attributable to the induction of another dielectric relaxation from space-charge polarization via ion transport (non-ideal cell geometry with finite conductivity of ITO electrodes). Because the flexoelectric switching is polar, the magnitude of the flexoelectric response of LC molecules would be a function of the frequency of the AC electric field. This allows us to divide the dielectric spectrum in Figure 7 into three frequency regimes by two designated frequencies of $f_{flexo} = 200$ Hz at $\varepsilon \equiv \varepsilon_L - 0.1\varepsilon_{flexo}$ and $f_{di} = 2500$ Hz at $\varepsilon \equiv \varepsilon_H + 0.1\varepsilon_{flexo}$ to characterize the frequency dependence of flexoelectric switching. As specified in [16], the flexoelectric effect is significant and independent of the frequency in region I ($f < f_{flexo} = 200$ Hz), but it is completely suppressed in region III ($f > f_{di} = 2500$ Hz).

Based on the frequency regions defined, optical properties of the cell in the PN-ULH state driven by voltages at $f = 100$ Hz in region I (Figure 8) or 5 kHz in region III (Figure 9) were discussed by measuring transmission spectra in the wavelength range of 400 nm–750 nm. According to Figure 7, the onset frequency for space-charge-polarization-induced dielectric relaxation is lower than 100 Hz. Our pre-tests ensured that no helical CLC textures could be electrically induced in the investigated frequency regime (100 Hz–5 kHz), such as Grandjean planar and focal-conic in general by the dielectric effect, and dynamic scattering, stripe, and Williams domains in particular via the ionic effect. Here, the cell was situated between a pair of crossed polarizers; thus, the voltage-dependent intensity of light propagating through the birefringent ULH texture can be written as:

$$I = I_0 \sin^2\{2[\alpha + \phi(V)]\} \sin^2\left\{\frac{\pi d}{\lambda}[n_{\mathrm{eff}}(V) - n_\mathrm{o}]\right\} \tag{2}$$

where I_0 is the intensity of incident light, α, as defined earlier in the Experimental section, is the angle between the transmission axis of a polarizer and the ULH helical axis at $V = 0$ V, λ is the wavelength of the incident light, $\phi(V)$ is the voltage-induced in-plane deviation angle of the ULH helical axis by the flexoelectric effect, and $n_{\mathrm{eff}}(V)$ as the effective refractive index is a function of the voltage-induced helical deformation by the dielectric effect. In the field-off state ($V = 0$ V), $\phi(V = 0\text{ V}) = 0^\circ$ and $n_{\mathrm{eff}}(V = 0\text{ V}) = (n_\mathrm{e} + n_\mathrm{o})/2$ so that the transmittance is zero at $\alpha = 0^\circ$ (Figure 8a) and the optical spectrum at $\alpha = 45^\circ$ exhibits wavelength dispersion of transmission with a peak transmittance of $T\%_{\mathrm{peak}}$ ~84%

at λ_{peark} ~650 nm and a minimal transmittance of $T\%_{\text{min}}$ ~0% at λ_{min} ~440 nm (Figure 9a). Accordingly, as supported by Equation (2), the magnitude of $T\%_{\text{peak}}$ would be primarily determined by ϕ (V) as long as the phase retardation in the second term is greater than $\pi/2$, and λ_{peark}, as well as λ_{min}, would be a function of n_{eff} (V). By setting $\alpha = 0°$ and applying a voltage at $f = 100$ Hz ($<f_{\text{flexo}} = 200$ Hz in region I) across the cell thickness, the optical transmission profiles as delineated in Figure 8a were strongly dependent on the voltage amplitude, showing increased $T\%_{\text{peak}}$ from ~3.7% at $V = 10$ V_{rms} to ~56% at $V = 70$ V_{rms} and, in the meantime, blue-shifted λ_{peark} and λ_{min} at $V > 20$ V_{rms}. This manifests a varying degree of splay-bend deformation by the flexoelectric effect and helical unwinding by the dielectric effect, induced by voltage at a frequency in region I, giving rise to the modulation of light intensity and phase retardation by changing ϕ (V) and n_{eff} (V) simultaneously. Notably, while plotting the voltage-dependent transmission $(V-T\%)$ curves with increasing and decreasing voltages from the optical spectral data, hysteresis-free electro-optic responses with a contrast ratio of ~80 (as calculated by the ratio between maximal transmittance of ~50.7% at 75 V_{rms} and minimal transmittance of ~0.6% at 0 V) were demonstrated (Figure 8b) thanks to the stabilization of the PN-ULH texture with high recoverability in the RM25TM25-CLC cell. The time in response to the voltage at $f = 100$ Hz at $T = 25$ °C, attributed to the combination of flexoelectric and dielectric effect, was on the order of sub-millisecond, varying from 1 ms at $V = 10$ V_{rms} to 0.28 ms at $V = 70$ V_{rms} (data not shown). On the other hand, when the voltage frequency was changed from 100 Hz to 5 kHz ($>f_{\text{di}} = 2500$ Hz in region III), it is clear from Figure 9a that the transmission spectra at $\alpha = 45°$ with nearly invariant $T\%_{\text{peak}}$ ~84% blue-shifted with ascending voltage amplitude beyond a threshold at ~20 V_{rms}. This suggests that the flexoelectric effect is considerably suppressed in the frequency region III, and the optical responses to the external voltage are primarily dominated by the dielectric effect and thereby the change of n_{eff} (V) with the voltage strength. The $V-T\%$ curves in Figure 9b denote that the electro-optic responses characterized by the dielectric effect at $f = 5$ kHz in region III were hysteresis-free as well, supporting again that the PN-ULH can be completely recovered after the CLC helix is partially or fully unwound by the applied voltage. Deduced from the results in Figure 9b, the threshold voltage for unwinding the helix was ~20 V_{rms}, and the voltage for sustaining LC molecules in the homeotropic state was ~100 V_{rms}.

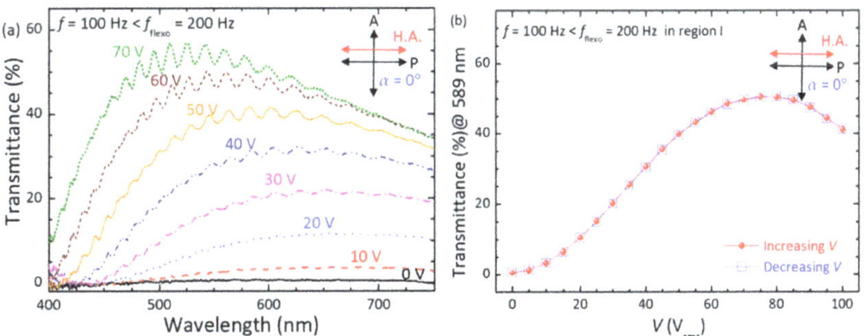

Figure 8. (**a**) Transmission spectra of the RM25TM25-CLC cell in the PN-ULH state driven by various voltages at a fixed frequency of 100 Hz. (**b**) Voltage-dependent transmission curves at $\lambda = 589$ nm and $f = 100$ Hz in the increasing- and decreasing-voltage processes. The cell for measurements at $T = 25$ °C was placed between crossed polarizers at a fixed polarizing angle of $\alpha = 0°$.

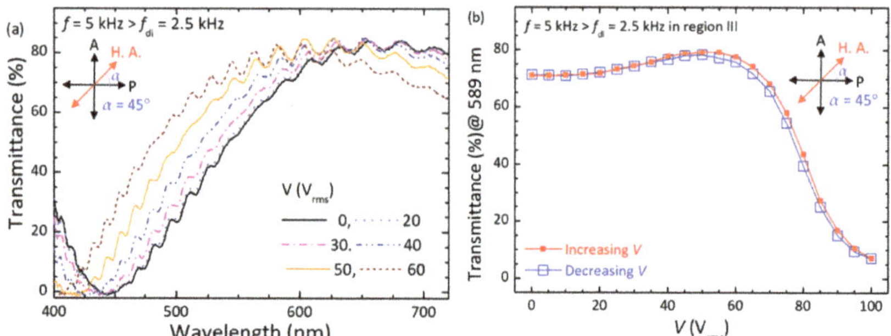

Figure 9. (**a**) transmission spectra of the RM25TM25-CLC cell in the PN-ULH state driven by various voltages at 5 kHz. (**b**) Voltage-dependent transmission curves at λ = 589 nm and f = 5 kHz upon increasing and decreasing voltage. The cell for measurements at T = 25 °C was placed between crossed polarizers at a fixed polarizing angle of α = 45°.

By fixing the voltage amplitude and permitting the frequency as a variable, an unusual frequency-modulated electro-optic feature, demonstrating electrically tunable light intensity without phase shift, was initiated for the first time to the best of our knowledge. This is evidenced by the results shown in Figure 10a using V = 40 V_{rms} as the example, which yielded invariant λ_{min} at ~425 nm and a gradual decrement in $T\%_{peak}$ with increasing frequency. It has been ensured that the intense frequency-dependent optical characteristics can also be realized at an arbitrary voltage in the range of 20 < V < 100 V_{rms} with a varying degree of tunable light intensity range (Figure 10b) but the value of λ_{min} as a constant blue-shifted with increasing voltage. The underlying mechanism can be explained by the frequency dependence of the flexoelectric effect and frequency-independent dielectric effect in the frequency regime featuring the flexoelectric-polarization-dominated dielectric relaxation. The dispersion of transmittance of f–$T\%$ curves in Figure 10b is analogous to that of the dielectric permittivity of the dielectric spectrum in Figure 7. Consequently, frequency-dependent electro-optic responses reported in this work are beneficial. This advantage enables the dual operation of the proposed bimesogen-doped CLC cell with a stable PN-ULH texture in the amplitude mode for controlling the light intensity without phase shift by the voltage *frequency* (based on the voltage-induced flexoelectric and dielectric effects) and in the phase mode for modulating the phase retardation by the voltage *amplitude* at a fixed frequency in region III (according to the dielectric effect). It is worth noting here that the dielectric relaxation and frequency-modulated electro-optic responses obtained in the PN-ULH are attributable to the CLC host with a significant flexoelectric effect due to the incorporation of CB7CB. Therefore, all the results demonstrated in this section would be enabled or reproducible by using other bent-core bimesogens if they can play the role of the main component in the explored CLC mixture to obtain the bimesogen-enhanced flexoelectric effect [28,35].

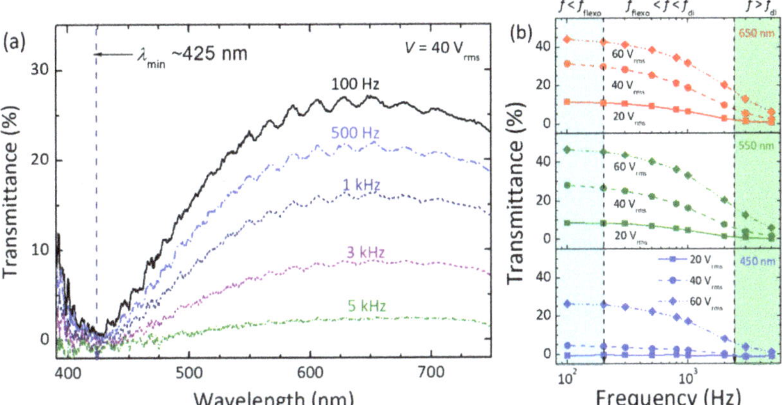

Figure 10. (**a**) Transmission spectra of the RM25TM25-CLC cell in the PN-ULH state driven by $V = 40~V_{rms}$ at various frequencies. (**b**) Frequency-dependent transmission at $\lambda = 450$ nm, 550 nm, and 650 nm of the cell applied with $V = 20~V_{rms}$, $40~V_{rms}$, and $60~V_{rms}$, respectively. Measurements were performed with crossed polarizers at $\alpha = 0°$.

4. Conclusions

We have produced a PN-ULH texture and explored its electro-optic characteristics in a bimesogen-doped CLC. With a fixed monomer concentration of 5-wt% incorporated into the CLC, our experimental results showed that the polymer network structure with 2.5-wt% RM257 and 2.5-wt% TMPTA in the RM25TM25-CLC cell is more sufficient than that with 5.0-wt% RM257 in the RM50-CLC for sustaining the desired ULH texture with high stability in the static state and excellent recoverability after cycles of electrically induced textural transition to the homeotropic state (Figures 3 and 4) and thermally induced phase transition to the isotropic phase (Figure 5). Evidence from the SEM images inferred that the copolymerization of RM257 and TMPTA promotes the generation of periodic polymer networks at both substrate surfaces with a periodicity of ~96 nm, which equals the half-pitch of the CLC (Figure 6).

Focusing on the PN-ULH texture in the RM25TM25-CLC, frequency-dependent flexo-electric effect, originating from the bimesogen-doped CLC host, was characterized by the real-part dielectric spectrum (Figure 7). In accordance with the profile of dielectric relaxation induced by the flexoelectric polarization, two frequency regions of $f < f_{\text{flexo}} = 200$ Hz (region I) and $f > f_{\text{di}} = 2500$ Hz (region III), as specified with the pronounced and suppressed flexoelectric effect, respectively, were defined to investigate the effect of voltage frequency on the electro-optic responses of the PN-ULH texture. Results of the voltage-dependent transmission spectra indicated that the light intensity is electrically tunable together with the blue-shift of the optical spectrum (transmission minimum in Figure 8a) with increasing voltage at $f = 100$ Hz in region I because of the induction of flexoelectric and dielectric effects. By varying the frequency to 5 kHz in region III, LC molecules were unable to respond to the electric field for flexoelectric switching so that only the phase-shift effect was obtained by the dielectric coupling of LCs with the voltage (Figure 9a). It is worth noting based on hysteresis-free $V-T$ curves in Figures 8b and 9b that, particularly due to the frequency-dependent flexoelectric effect and frequency-independent dielectric effect in the frequency range highlighted by the flexoelectric-polarization-induced dielectric relaxation, a superior feature, enabling frequency modulation of light intensity without phase shift, was implemented (Figure 10). Accordingly, the proposed bimesogen-doped CLC, featuring the highly stable ULH texture by polymer stabilization and the alluring function of frequency-modulated electro-optic characteristics, paves a new pathway toward developing a dual (amplitude and phase modulation)-mode electro-optic device for potential applications in information displays, phase retarders, spatial light modulators, and other optical and photonic devices. For the proposed bimesogen-doped CLC system, miscible bent-core and rod-like LCs are typically polar in nature, and a strong flexoelectric effect is induced generally by low-frequency voltages. Further discussion on the issue of the impact of ionic effect on the electrical and electro-optic characteristics (e.g., the threshold voltage, hysteresis, response time, and voltage holding ratio) is essential for advancing technical applications.

Author Contributions: Conceptualization, P.-C.W.; methodology, C.-H.Y. and P.-C.W.; software, C.-H.Y. and P.-C.W.; validation, C.-H.Y., P.-C.W. and W.L.; formal analysis, C.-H.Y. and P.-C.W.; resources, W.L.; data curation, C.-H.Y. and P.-C.W.; writing—original draft preparation, C.-H.Y.; writing—review and editing, P.-C.W. and W.L.; visualization, C.-H.Y. and P.-C.W.; supervision, W.L.; project administration, W.L.; funding acquisition, W.L. All authors have read and agreed to the published version of the manuscript.

Funding: This research was funded by the Ministry of Science and Technology, Taiwan, grant number 110-2112-M-A49-023.

Institutional Review Board Statement: Not applicable.

Informed Consent Statement: Not applicable.

Data Availability Statement: The authors confirm that the data supporting the findings of this study are available within the article.

Conflicts of Interest: The authors declare no conflict of interest.

References

1. Rudquist, P.; Carlsson, T.; Komitov, L.; Lagerwall, S. The flexoelectro-optic effect in cholesterics. *Liq. Cryst.* **1997**, *22*, 445–449. [CrossRef]
2. Rudquist, P.; Komitov, L.; Lagerwall, S. Volume-stabilized ULH structure for the flexoelectro-optic effect and the phase-shift effect in cholesterics. *Liq. Cryst.* **1998**, *24*, 329–334. [CrossRef]
3. Patel, J.; Meyer, R.B. Flexoelectric electro-optics of a cholesteric liquid crystal. *Phys. Rev. Lett.* **1987**, *58*, 1538. [CrossRef] [PubMed]
4. Tan, G.; Lee, Y.-H.; Gou, F.; Chen, H.; Huang, Y.; Lan, Y.-F.; Tsai, C.-Y.; Wu, S.-T. Review on polymer-stabilized short-pitch cholesteric liquid crystal displays. *J. Phys. D Appl. Phys.* **2017**, *50*, 493001. [CrossRef]
5. Fells, J.A.; Welch, C.; Yip, W.C.; Elston, S.J.; Booth, M.J.; Mehl, G.H.; Wilkinson, T.D.; Morris, S.M. Dynamic response of large tilt-angle flexoelectro-optic liquid crystal modulators. *Opt. Express* **2019**, *27*, 15184–15193. [CrossRef] [PubMed]
6. Yip, W.; Welch, C.; Mehl, G.H.; Wilkinson, T.D. Analog modulation by the flexoelectric effect in liquid crystals. *Appl. Opt.* **2020**, *59*, 2668–2673. [CrossRef]
7. Wang, X.; Fells, J.A.; Shi, Y.; Ali, T.; Welch, C.; Mehl, G.H.; Wilkinson, T.D.; Booth, M.J.; Morris, S.M.; Elston, S.J. A compact full 2π flexoelectro-optic liquid crystal phase modulator. *Adv. Mater. Technol.* **2020**, *5*, 2000589. [CrossRef]
8. Hyman, R.M.; Lorenz, A.; Wilkinson, T.D. Phase modulation using different orientations of a chiral nematic in liquid crystal over silicon devices. *Liq. Cryst.* **2016**, *43*, 83–90. [CrossRef]
9. Yoshida, H.; Inoue, Y.; Isomura, T.; Matsuhisa, Y.; Fujii, A.; Ozaki, M. Position sensitive, continuous wavelength tunable laser based on photopolymerizable cholesteric liquid crystals with an in-plane helix alignment. *Appl. Phys. Lett.* **2009**, *94*, 72. [CrossRef]
10. Coles, H.; Clarke, M.; Morris, S.; Broughton, B.; Blatch, A. Strong flexoelectric behavior in bimesogenic liquid crystals. *J. Appl. Phys.* **2006**, *99*, 034104. [CrossRef]
11. Morris, S.; Clarke, M.; Blatch, A.; Coles, H. Structure-flexoelastic properties of bimesogenic liquid crystals. *Phys. Rev. E* **2007**, *75*, 041701. [CrossRef]
12. Hegde, G.; Komitov, L. Periodic anchoring condition for alignment of a short pitch cholesteric liquid crystal in uniform lying helix texture. *Appl. Phys. Lett.* **2010**, *96*, 113503. [CrossRef]
13. Komitov, L.; Bryan-Brown, G.; Wood, E.; Smout, A. Alignment of cholesteric liquid crystals using periodic anchoring. *J. Appl. Phys.* **1999**, *86*, 3508–3511. [CrossRef]
14. Wang, C.-T.; Wang, W.-Y.; Lin, T.-H. A stable and switchable uniform lying helix structure in cholesteric liquid crystals. *Appl. Phys. Lett.* **2011**, *99*, 041108. [CrossRef]
15. Nian, Y.-L.; Wu, P.-C.; Lee, W. Optimized frequency regime for the electrohydrodynamic induction of a uniformly lying helix structure. *Photonics Res.* **2016**, *4*, 227–232. [CrossRef]
16. Lin, Y.-C.; Wu, P.-C.; Lee, W. Frequency-modulated textural formation and optical properties of a binary rod-like/bent-core cholesteric liquid crystal. *Photonics Res.* **2019**, *7*, 1258–1265. [CrossRef]
17. Yu, C.-H.; Wu, P.-C.; Lee, W. Alternative generation of well-aligned uniform lying helix texture in a cholesteric liquid crystal cell. *AIP Adv.* **2017**, *7*, 105107. [CrossRef]

18. Yu, C.-H.; Wu, P.-C.; Lee, W. Electro-thermal formation of uniform lying helix alignment in a cholesteric liquid crystal cell. *Crystals* **2019**, *9*, 183. [CrossRef]

19. Gardiner, D.J.; Morris, S.M.; Hands, P.J.; Castles, F.; Qasim, M.M.; Kim, W.-S.; Seok Choi, S.; Wilkinson, T.D.; Coles, H.J. Spontaneous induction of the uniform lying helix alignment in bimesogenic liquid crystals for the flexoelectro-optic effect. *Appl. Phys. Lett.* **2012**, *100*, 063501. [CrossRef]

20. Li, C.-C.; Tseng, H.-Y.; Chen, C.-W.; Wang, C.-T.; Jau, H.-C.; Wu, Y.-C.; Hsu, W.-H.; Lin, T.-H. Versatile energy-saving smart glass based on tristable cholesteric liquid crystals. *ACS Appl. Energy Mater.* **2020**, *3*, 7601–7609. [CrossRef]

21. Varanytsia, A.; Chien, L.-C. Giant flexoelectro-optic effect with liquid crystal dimer CB7CB. *Sci. Rep.* **2017**, *7*, 1–7. [CrossRef]

22. Varanytsia, A.; Chien, L.-C. Bimesogen-enhanced flexoelectro-optic behavior of polymer stabilized cholesteric liquid crystal. *J. Appl. Phys.* **2016**, *119*, 014502. [CrossRef]

23. Kim, S.H.; Chien, L.-C.; Komitov, L. Short pitch cholesteric electro-optical device stabilized by nonuniform polymer network. *Appl. Phys. Lett.* **2005**, *86*, 161118. [CrossRef]

24. Broughton, B.; Clarke, M.; Morris, S.; Blatch, A.; Coles, H. Effect of polymer concentration on stabilized large-tilt-angle flexoelectro-optic switching. *J. Appl. Phys.* **2006**, *99*, 023511. [CrossRef]

25. Tartan, C.C.; Salter, P.S.; Booth, M.J.; Morris, S.M.; Elston, S.J. Localised polymer networks in chiral nematic liquid crystals for high speed photonic switching. *J. Appl. Phys.* **2016**, *119*, 183106. [CrossRef]

26. Broughton, B.; Clarke, M.; Blatch, A.; Coles, H. Optimized flexoelectric response in a chiral liquid-crystal phase device. *J. Appl. Phys.* **2005**, *98*, 034109. [CrossRef]

27. Castles, F.; Morris, S.; Coles, H. Flexoelectro-optic properties of chiral nematic liquid crystals in the uniform standing helix configuration. *Phys. Rev. E* **2009**, *80*, 031709. [CrossRef] [PubMed]

28. Outram, B.; Elston, S. Frequency-dependent dielectric contribution of flexoelectricity allowing control of state switching in helicoidal liquid crystals. *Phys. Rev. E* **2013**, *88*, 012506. [CrossRef]

29. Cestari, M.; Diez-Berart, S.; Dunmur, D.; Ferrarini, A.; De La Fuente, M.; Jackson, D.; Lopez, D.; Luckhurst, G.; Perez-Jubindo, M.; Richardson, R. Phase behavior and properties of the liquid-crystal dimer 1″, 7′-bis (4-cyanobiphenyl-4′-yl) heptane: A twist-bend nematic liquid crystal. *Phys. Rev. E* **2011**, *84*, 031704. [CrossRef]

30. Xiang, J.; Li, Y.; Li, Q.; Paterson, D.A.; Storey, J.M.; Imrie, C.T.; Lavrentovich, O.D. Electrically tunable selective reflection of light from ultraviolet to visible and infrared by heliconical cholesterics. *Adv. Mater.* **2015**, *27*, 3014–3018. [CrossRef]

31. Hou, H.; Gan, Y.; Yin, J.; Jiang, X. Polymerization-induced growth of microprotuberance on the photocuring coating. *Langmuir* **2017**, *33*, 2027–2032. [CrossRef] [PubMed]

32. Gao, L.; Wang, K.-M.; Zhao, R.; Ma, H.-M.; Sun, Y.-B. Effect of a dual functional polymer on the electro-optical properties of blue phase liquid crystals. *Polymers* **2019**, *11*, 1128. [CrossRef] [PubMed]

33. Chen, D.; Nakata, M.; Shao, R.; Tuchband, M.R.; Shuai, M.; Baumeister, U.; Weissflog, W.; Walba, D.M.; Glaser, M.A.; Maclennan, J.E. Twist-bend heliconical chiral nematic liquid crystal phase of an achiral rigid bent-core mesogen. *Phys. Rev. E* **2014**, *89*, 022506. [CrossRef] [PubMed]

34. Yu, M.; Zhou, X.; Jiang, J.; Yang, H.; Yang, D.-K. Matched elastic constants for a perfect helical planar state and a fast switching time in chiral nematic liquid crystals. *Soft Matter* **2016**, *12*, 4483–4488. [CrossRef] [PubMed]

35. Tan, G.; Lee, Y.-H.; Gou, F.; Hu, M.; Lan, Y.-F.; Tsai, C.-Y.; Wu, S.-T. Macroscopic model for analyzing the electro-optics of uniform lying helix cholesteric liquid crystals. *J. Appl. Phys.* **2017**, *121*, 173102. [CrossRef]

Article

Double-Resolved Beam Steering by Metagrating-Based Tamm Plasmon Polariton

Rashid G. Bikbaev [1,2,*], Dmitrii N. Maksimov [1,2], Kuo-Ping Chen [3,4] and Ivan V. Timofeev [1,2]

1   Kirensky Institute of Physics, Federal Research Center KSC SB RAS, Krasnoyarsk 660036, Russia
2   Siberian Federal University, Krasnoyarsk 660041, Russia
3   Institute of Photonics Technologies, National Tsing Hua University, Hsinchu 30013, Taiwan
4   Institute of Imaging and Biomedical Photonics, National Yang Ming Chiao Tung University, Tainan 71150, Taiwan
*   Correspondence: bikbaev@iph.krasn.ru

Abstract: We consider Tamm plasmon polariton in a subwavelength grating patterned on top of a Bragg reflector. We demonstrate dynamic control of the phase and amplitude of a plane wave reflected from such metagrating due to resonant coupling with the Tamm plasmon polariton. The tunability of the phase and amplitude of the reflected wave arises from modulation of the refractive index of a transparent conductive oxide layer by applying the bias voltage. The electrical switching of diffracted beams of the ±1st order is shown. The possibility of doubling the angular resolution of beam steering by using asymmetric reflected phase distribution with integer and half-integer periods of the metagrating is demonstrated.

Keywords: metasurface; metagratings; tamm plasmon polaritons; chirality

1. Introduction

The metasurface is an artificially structured layer with subwavelength-scaled elements designed to effectively control the phase and amplitude of reflected light [1,2]. The interest in such systems is fueled by a significant advance in subwavelength technologies such as optical and e-beam lithography. Metasurfaces open up new opportunities for implementation of holograms [3], lenses [4], media with anomalous reflection [5,6], lasers [7,8], perfect absorber with critical coupling [9], etc. Structures in which meta-atoms or metamolecules form two-dimensional periodic lattices are often called metagratings [10,11]. Such constructions have the ability to direct light into a certain diffraction channel, achieving remarkable efficiency even at very large bending angles [6,12]. An actively tuned metasurface and metagrating with control of phase and amplitude of individual elements ensures the generation of an arbitrarily complex wave front. There are several approaches to actively rearrange the optical properties of a metasurface and metagrating. The first one is application of liquid crystals [13–15]. The advantage of this method is the control over optical properties via variation of both electric field and temperature. However, one of the key drawbacks of such devices is their large switching times of several milliseconds. Another promising approach is application of phase-change metasurfaces and metagrating. These systems have proven to have excellent switching times of microseconds, but the phase modulation schemes applied so far only allowed for discrete switching [16]. Variation of the external electric field has already been implemented in plasmonic amplitude modulators [17,18]. It is shown that classical semiconductor materials can be replaced by transparent conducting oxides. Thus, the effective control of the volume concentration of charge carriers at the interface between a conducting oxide and a metal film was demonstrated [19,20].

It was shown that the increase in the applied bias voltage leads to a zero and even negative real part of the permittivity of the conducting oxide in a very thin layer near

Citation: Bikbaev, R.G.; Maksimov, D.N.; Chen, K.-P.; Timofeev, I.V. Double-Resolved Beam Steering by Metagrating-Based Tamm Plasmon Polariton. *Materials* **2022**, *15*, 6014. https://doi.org/10.3390/ma15176014

Academic Editor: George Kenanakis

Received: 12 August 2022
Accepted: 29 August 2022
Published: 31 August 2022

Publisher's Note: MDPI stays neutral with regard to jurisdictional claims in published maps and institutional affiliations.

the metal. This results in a significant field enhancement in this region and provides the necessary phase jump near the resonant wavelength.

In this work, we demonstrate effective tuning of the wavefront in the near infrared region by introducing a controlled material into a structure (Figure 1a) supporting Tamm plasmon polaritons (TTP) [21–25]. The beam steering in indium tin oxide (ITO) [26,27] metasurfaces has been previously demonstrated in [20]. Here we show that our scheme allows for doubling the angular resolution of beam steering with phase distributions with integer and half-integer periods of the metagrating. The advantage of the proposed TTP-based structure compared to conventional metal–insulator–metal devices is the excitation of resonance with a higher Q-factor which makes the phase more sensitive to the control parameters. Moreover, the proposed structure is attractive for implementing photonic crystal surface-emitting lasers (PCSEL) [28]. Thus, the setup proposed paves the way for engineering optical devices with both PCSEL and beam steering functionality.

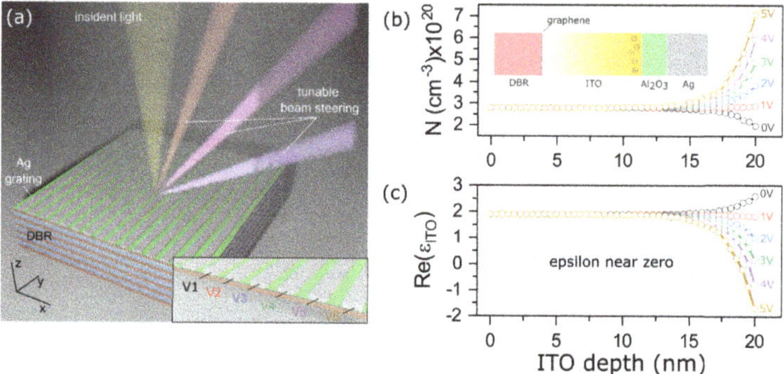

Figure 1. (**a**) Sketch view of the structure; (**b**,**c**) electron concentration N and real part of the dielectric permittivity $\mathrm{Re}\varepsilon_{ITO}$ of the ITO layer for different applied bias voltage. The DBR layers' thicknesses are $d_a = 165$ nm and $d_b = 135$ nm, for silica and titanium dioxide, correspondingly. The number of DBR layers are equal to 15. The 2D array with thickness $h = 95$ nm and width $L = 470$ nm has infinite length along the y axis. The pitch of the array $p = 500$ nm. The ITO and Al_2O_3 layer thicknesses are 20 nm and 5 nm, respectively. The structure in the inset of Figure 1b is presented schematically to demonstrate the non-uniform distribution of charges in the ITO in the case of applying a bias voltage between Ag nanostripes and monolayer graphene.

2. Model

2.1. Electrostatic Simulation

The electron distribution in the ITO layer as a function of applied bias voltage was define by numerically solving the Poisson and drift-diffusion equations. The ITO has been presented as a semiconductor with the bandgap of $E = 2.8$ eV, electron affinity of $\chi = 5$ eV, effective electron mass of $m^* = 0.25m_e$ and permittivity $\varepsilon = 9$. The ITO carrier concentration is $N_0 = 2.8 \times 10^{20}$ cm^{-3}. The DC permittivity of the Al_2O_3 is $\varepsilon_{Al_2O_3} = 9$. In our simulation, the Ag nanostripes and a monolayer graphene were used as electric contacts. The change in the dielectric constant of Ag under the bias voltage was not taken into account. We used the mesh size of 0.1 nm which has been validated by performing careful convergence tests.

It can be seen from Figure 1b that at zero voltage, electron depletion is observed at the ITO-Al_2O_3 boundary. This is due to the fact that the working function of the ITO (4.4 eV) at this concentration is lower than the working function of silver (4.73 eV). The increase of the bias voltage between nanostripes and graphene layer leads to electron accumulation at the ITO-Al_2O_3 interface. Figure 1c shows the dependence of the real part of the complex permittivity of the ITO layer at a wavelength of 1550 nm on the applied bias voltage and the distance from the Al_2O_3 boundary. The increase in voltage leads to a significant change in the dielectric constant of the conductive oxide in a thin 3 nm layer. At voltages greater than 2 V, it takes values close to zero. The epsilon near zero region area is highlighted with blue. A further increase in voltage leads to the real part of the dielectric constant becoming negative.

2.2. Full-Wave Simulation

The reflection spectra of the structure from Figure 1a and the reflection phases for normally incident TM-polarized waves were calculated by the finite-difference time domain (FDTD) method. The DBR unit cell is formed by silica and titanium dioxides. The geometric parameters of the DBR are listed in the caption of Figure 1a.

The reflectance spectra of the structure at different values of the bias voltage applied to the ITO are shown in Figure 2a. The dip observed near $\lambda = 1550$ nm in Figure 2a corresponds to the Tamm plasmon polariton localized at the interface between the metagrating and the multilayer mirror [29]. The minimum reflection at the TPP wavelength is due to the critical couplin, at which the rate of energy decay into the radiation and absorption channels of the metagrating are comparable. In practice, this effect is achieved by increasing the number of layers of the multilayer mirror and tuning the height of silver nanostrips. An increase in the bias voltage leads to a blue shift of the resonant wavelength. This effect can be explained by the fact that the increase in voltage leads to an increase in the volume concentration of the charge carriers near the ITO-Al_2O_3 boundary. As a result, the real part of the complex dielectric permittivity of the ITO becomes negative, and it acquires metallic properties. Thus, an additional term appears in the expression of phase matching corresponding to the phase winding when passing through the ITO layer. The blue shift of the localized state leads to a significant change in the reflection phase at a wavelength of $\lambda = 1550$ nm (see Figure 2b). Thus, at a voltage of 3 V, it is possible to achieve a phase change of 180°, while at the maximum considered voltage of 5 V, the phase is 214°.

The calculations have shown that by changing the bias voltage applied to the ITO film, it is possible to control the reflection phase of each nanostrip. Thus, it becomes possible to create a tunable diffraction grating, the period of which is determined by the number of nanostrips with different applied bias voltages. The angular resolution of such a device is limited by the width of the unit cell, since switching is carried out discretely, see the diagram in Figure 2c. So, in the case when a voltage of 0 V is applied to two strips, and 3.5 V to the next two strips, the lattice period is two microns. Changing the number of strips from four to eight leads to an increase of the grating period to four microns and, as a consequence, results in a change of the first-order diffraction angle:

$$\Theta_n = -\arcsin\left(\frac{m\lambda}{np}\right) \tag{1}$$

here and after the diffraction order $m = 0, \pm1$, and n is the number of strips used in the dynamic unit cell. In principle, n can be an arbitrary real number providing continuous steering rather than discrete steering for integer n. The drawback of non-integer n is that the TPPs of neighboring strips are mixed; their phases produce unrepeatable cell patterns that are hard to control. Nevertheless, rational n produces few different patterns that are periodic in space. For example, the half-integer value of n makes only two different cell patterns that lead to relatively sharp beams, as illustrated in Table 1. If three and more patterns are included, then the ±1 diffraction order intensity falls down and requires special inverse design of applied voltages for effective beam steering.

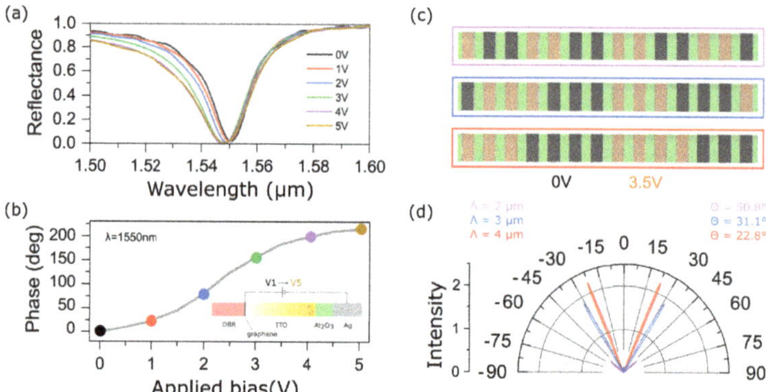

Figure 2. (**a,b**) Reflectance spectra of the structure presented in Figure 1a,b simulated phase shift as a function of applied bias voltage between Ag nanostripes and monolayer graphene; (**c**) schematic of the diffraction grating for a different number of nanostrip pairs; dark gray and orange nanostrips depict nanostrips without bias voltage and with bias voltage of 3.5 V, respectively; an increase in the number of nanostripes with bias voltage and without it leads to an increase in the grating period; (**d**) Simulated far-field reflected intensity from the metagrating as a function of the diffraction angles for a different grating period.

To numerically demonstrate this effect, we calculated the intensities of diffraction maxima in the far field. The calculation was carried out for 50 nanostrips. The results are presented in Figure 2d. It can be seen that a change in the lattice period leads to a significant (about 30°) change in the angles of −1 and +1 diffraction orders. Note that the intensity of zeroth order is equal to zero. This is explained by the destructive interference of waves propagating along the normal to the metagrating due to the fact that the phase difference of the reflected waves from neighboring nanostrips is equal to π.

Table 1. The dependence of the ± first order diffraction angle on number of nanostrips. Half-integer n corresponds to double resolution.

Number of Strips, n	Θ_n Simulated by FDTD	Θ_n Calculated by Equation (1)	Results
3	>90	>90	
3.5	62.3	62.33	Figure 3c
4	50.77	50.80	Figure 2d
4.5	-	43.54	
5	38.3	38.32	Figure 3c
5.5	-	34.30	
6	31.10	31.10	Figure 2d
6.5	-	28.48	
7	-	26.28	
7.5	-	24.41	
8	22.8	22.79	Figure 2d

We aim at doubling the angle resolution of the device by continuously tuning the voltage and, as a consequence, the phase from one nanostrip to another, see Figure 3a. For double resolution, one can add a diffraction grating period consisting of a half-integer number of nanostrips along with integer nanostrip periods, see Figure 3b. In this regard, it becomes relevant to solve the inverse problem, i.e., to calculate the phase distribution over nanostrips and to provide reflection at the necessary angle via the generalized Snell's law:

$$\sin(\theta_r) - \sin(\theta_i) = \frac{\lambda}{2\pi} \frac{\Delta\Phi}{p},$$

(2)

where $\Delta\Phi$ is the phase difference between neighbouring nanostrips, and θ_r and θ_i are reflection and incidence angles, respectively. To demonstrate this effect, the required phase distribution along the metagrating was determined, see Figure 3b. The calculations were performed for 50 nanostrips and for $\theta = 38.3°$ and $\theta = 62.3°$. It can be seen from Figure 3b that for an angle of $38.3°$ the diffraction grating period is formed of five nanostrips, while the phase distribution turns out to be asymmetric. For larger angles, for example for $62.3°$, the phase distribution is also asymmetric, but the period of the phase distribution consists of a half-integer number of nanostrips and is equal to 3.5, see Table 1.

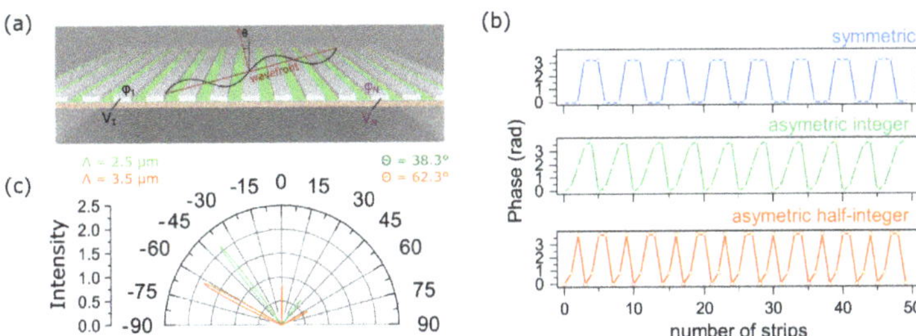

Figure 3. (**a**) Schematic representation of the structure for control over the angle of the first diffraction order; (**b**) three types of the phase distribution along the metagrating; (**c**) simulated far-field reflected intensity from the metagrating based on phase distribution presented in (**b**) in the case of asymmetric phase distribution.

Based on this distribution, the corresponding bias voltages were determined by interpolating the data presented in Figure 2b. According to the obtained results, the dependence of the refractive index of the ITO film on its thickness was determined. Then, the intensities of diffraction maxima in the far field were calculated. The results are shown in Figure 3b. It can be seen from Figure 3b that the intensity of -1 order is much greater than the intensity of $+1$ order. This is explained by the fact that the phase profile shown in Figure 3b leads to an increase in the intensity of the field only for -1 order, while for 0 and $+1$ order direction, the waves are destructively extinguished. Thus, the designed scheme allows one to control a nanoscale beam in a wide range of angles and has significant potential for engineering ultrathin devices such as LIDARs and nanoscale spatial light modulators.

3. Conclusions

The paper demonstrates the control of the phase and amplitude of a wave reflected from a Tamm plasmon polariton structure. The refractive index modulation in a thin layer of transparent conductive oxide located at the boundary of a multilayer mirror and a grating metagrating on the phase of the reflected wave is demonstrated. This effect makes it possible to use such a structure as a dynamic phase diffraction grating. The calculations showed that changing the number of nanostrips with different applied bias voltage allows for efficient switching in $\pm$ first diffraction orders. In addition, it is shown that the reflected beam can be controlled by a continuous phase change along the metagrating. This method allows us to increase the beam steering angular resolution.

Author Contributions: Conceptualization, R.G.B. and I.V.T.; methodology and software, R.G.B. and I.V.T.; validation, R.G.B., D.N.M., K.-P.C. and I.V.T.; writing—original draft preparation, R.G.B. and D.N.M.; visualization, R.G.B.; supervision, K.-P.C. and I.V.T. All the authors discussed the results, commented on the manuscript, and agreed to the published version of the manuscript. All authors have read and agreed to the published version of the manuscript.

Funding: This research was funded by the Russian Science Foundation (project no. 22-42-08003). This work was supported by the Higher Education Sprout Project of the National Yang Ming Chiao Tung University and Ministry of Education and the Ministry of Science and Technology (111-2923-E-A49-001-MY3, 110-2221-E-A49-019-MY3, 109-2628-E-009-007-MY3).

Institutional Review Board Statement: Not applicable.

Informed Consent Statement: Not applicable.

Data Availability Statement: The data presented in this study are available upon reasonable request from the corresponding author.

Conflicts of Interest: The authors declare no conflict of interest.

References

1. He, Q.; Sun, S.; Zhou, L. Tunable/Reconfigurable Metasurfaces: Physics and Applications. *Research* **2019**, *2019*, 1–16. [CrossRef] [PubMed]
2. Chen, W.T.; Zhu, A.Y.; Capasso, F. Flat optics with dispersion-engineered metasurfaces. *Nat. Rev. Mater.* **2020**, *5*, 604–620. [CrossRef]
3. Huang, L.; Zhang, S.; Zentgraf, T. Metasurface holography: from fundamentals to applications. *Nanophotonics* **2018**, *7*, 1169–1190. [CrossRef]
4. Bosch, M.; Shcherbakov, M.R.; Won, K.; Lee, H.S.; Kim, Y.; Shvets, G. Electrically Actuated Varifocal Lens Based on Liquid-Crystal-Embedded Dielectric Metasurfaces. *Nano Lett.* **2021**, *21*, 3849–3856. [CrossRef]
5. Sun, S.; Yang, K.Y.; Wang, C.M.; Juan, T.K.; Chen, W.T.; Liao, C.Y.; He, Q.; Xiao, S.; Kung, W.T.; Guo, G.Y.; et al. High-Efficiency Broadband Anomalous Reflection by Gradient Meta-Surfaces. *Nano Lett.* **2012**, *12*, 6223–6229. [CrossRef] [PubMed]
6. Li, Z.; Palacios, E.; Butun, S.; Aydin, K. Visible-Frequency Metasurfaces for Broadband Anomalous Reflection and High-Efficiency Spectrum Splitting. *Nano Lett.* **2015**, *15*, 1615–1621. [CrossRef]
7. Sroor, H.; Huang, Y.W.; Sephton, B.; Naidoo, D.; Vallés, A.; Ginis, V.; Qiu, C.W.; Ambrosio, A.; Capasso, F.; Forbes, A. High-purity orbital angular momentum states from a visible metasurface laser. *Nat. Photonics* **2020**, *14*, 498–503. [CrossRef]
8. Xu, W.H.; Chou, Y.H.; Yang, Z.Y.; Liu, Y.Y.; Yu, M.W.; Huang, C.H.; Chang, C.T.; Huang, C.Y.; Lu, T.C.; Lin, T.R.; et al. Tamm Plasmon-Polariton Ultraviolet Lasers. *Adv. Photonics Res.* **2021**, *3*, 2100120. [CrossRef]
9. Liang, Y.; Koshelev, K.; Zhang, F.; Lin, H.; Lin, S.; Wu, J.; Jia, B.; Kivshar, Y. Bound States in the Continuum in Anisotropic Plasmonic Metasurfaces. *Nano Lett.* **2020**, *20*, 6351–6356. [CrossRef]
10. Ra'di, Y.; Sounas, D.L.; Alù, A. Metagratings: Beyond the Limits of Graded Metasurfaces for Wave Front Control. *Phys. Rev. Lett.* **2017**, *119*, 067404. [CrossRef]
11. Panagiotidis, E.; Almpanis, E.; Stefanou, N.; Papanikolaou, N. Multipolar interactions in Si sphere metagratings. *J. Appl. Phys.* **2020**, *128*, 093103. [CrossRef]
12. Khorasaninejad, M.; Capasso, F. Broadband Multifunctional Efficient Meta-Gratings Based on Dielectric Waveguide Phase Shifters. *Nano Lett.* **2015**, *15*, 6709–6715. [CrossRef]
13. Li, J.; Yu, P.; Zhang, S.; Liu, N. Electrically-controlled digital metasurface device for light projection displays. *Nat. Commun.* **2020**, *11*, 3574. [CrossRef] [PubMed]
14. Su, H.; Wang, H.; Zhao, H.; Xue, T.; Zhang, J. Liquid-Crystal-Based Electrically Tuned Electromagnetically Induced Transparency Metasurface Switch. *Sci. Rep.* **2017**, *7*, 17378. [CrossRef]
15. Chen, K.P.; Ye, S.C.; Yang, C.Y.; Yang, Z.H.; Lee, W.; Sun, M.G. Electrically tunable transmission of gold binary-grating metasurfaces integrated with liquid crystals. *Opt. Express* **2016**, *24*, 16815. [CrossRef] [PubMed]
16. Cueff, S.; John, J.; Zhang, Z.; Parra, J.; Sun, J.; Orobtchouk, R.; Ramanathan, S.; Sanchis, P. VO$_2$ nanophotonics. *APL Photonics* **2020**, *5*, 110901. [CrossRef]
17. Lee, H.W.; Papadakis, G.; Burgos, S.P.; Chander, K.; Kriesch, A.; Pala, R.; Peschel, U.; Atwater, H.A. Nanoscale Conducting Oxide PlasMOStor. *Nano Lett.* **2014**, *14*, 6463–6468. [CrossRef]
18. Babicheva, V.E.; Boltasseva, A.; Lavrinenko, A.V. Transparent conducting oxides for electro-optical plasmonic modulators. *Nanophotonics* **2015**, *4*, 165–185. [CrossRef]
19. Lee, Y.; Yun, J.; Kim, S.J.; Seo, M.; In, S.; Jeong, H.D.; Lee, S.Y.; Park, N.; Chung, T.D.; Lee, B. High-Speed Transmission Control in Gate-Tunable Metasurfaces Using Hybrid Plasmonic Waveguide Mode. *Adv. Opt. Mater.* **2020**, *8*, 2001256. [CrossRef]
20. Huang, Y.W.; Lee, H.W.H.; Sokhoyan, R.; Pala, R.A.; Thyagarajan, K.; Han, S.; Tsai, D.P.; Atwater, H.A. Gate-Tunable Conducting Oxide Metasurfaces. *Nano Lett.* **2016**, *16*, 5319–5325. [CrossRef]
21. Kaliteevski, M.; Iorsh, I.; Brand, S.; Abram, R.A.; Chamberlain, J.M.; Kavokin, A.V.; Shelykh, I.A. Tamm plasmon-polaritons: Possible electromagnetic states at the interface of a metal and a dielectric Bragg mirror. *Phys. Rev. B* **2007**, *76*, 165415. [CrossRef]
22. Sasin, M.E.; Seisyan, R.P.; Kaliteevski, M.; Brand, S.; Abram, R.A.; Chamberlain, J.M.; Egorov, A.Y.; Vasil'ev, A.P.; Mikhrin, V.S.; Kavokin, A.V. Tamm plasmon polaritons: Slow and spatially compact light. *Appl. Phys. Lett.* **2008**, *92*, 251112. [CrossRef]
23. Bikbaev, R.; Vetrov, S.; Timofeev, I. Epsilon-Near-Zero Absorber by Tamm Plasmon Polariton. *Photonics* **2019**, *6*, 28. [CrossRef]
24. Vyunishev, A.M.; Bikbaev, R.G.; Svyakhovskiy, S.E.; Timofeev, I.V.; Pankin, P.S.; Evlashin, S.A.; Vetrov, S.Y.; Myslivets, S.A.; Arkhipkin, V.G. Broadband Tamm plasmon polariton. *J. Opt. Soc. Am. B* **2019**, *36*, 2299–2305. [CrossRef]
25. Wu, F.; Wu, X.; Xiao, S.; Liu, G.; Li, H. Broadband wide-angle multilayer absorber based on a broadband omnidirectional optical Tamm state. *Opt. Express* **2021**, *29*, 23976. [CrossRef]

26. Bohn, J.; Luk, T.S.; Tollerton, C.; Hutchings, S.W.; Brener, I.; Horsley, S.; Barnes, W.L.; Hendry, E. All-optical switching of an epsilon-near-zero plasmon resonance in indium tin oxide. *Nat. Commun.* **2021**, *12*, 1017. [CrossRef] [PubMed]
27. Pang, K.; Alam, M.Z.; Zhou, Y.; Liu, C.; Reshef, O.; Manukyan, K.; Voegtle, M.; Pennathur, A.; Tseng, C.; Su, X.; et al. Adiabatic Frequency Conversion Using a Time-Varying Epsilon-Near-Zero Metasurface. *Nano Lett.* **2021**, *21*, 5907–5913. [CrossRef]
28. Chen, L.R.; Chang, C.J.; Hong, K.B.; Weng, W.C.; Chuang, B.H.; Huang, Y.W.; Lu, T.C. Static Beam Steering by Applying Metasurfaces on Photonic-Crystal Surface-Emitting Lasers. *J. Light. Technol.* **2022**, 1–6. [CrossRef]
29. Bikbaev, R.G.; Maksimov, D.N.; Pankin, P.S.; Chen, K.P.; Timofeev, I.V. Critical coupling vortex with grating-induced high Q-factor optical Tamm states. *Opt. Express* **2021**, *29*, 4672–4680. [CrossRef]

 *materials* 

Article

Influence of Rugate Filters on the Spectral Manifestation of Tamm Plasmon Polaritons

Victor Yu. Reshetnyak [1], Igor P. Pinkevych [1,*], Timothy J. Bunning [2] and Dean R. Evans [2]

[1] Physics Faculty, Taras Shevchenko National University of Kyiv, 01601 Kyiv, Ukraine; VReshetnyak@univ.kiev.ua
[2] Air Force Research Laboratory, Materials and Manufacturing Directorate, Wright-Patterson Air Force Base, OH 45433, USA
* Correspondence: ipinkevych@univ.kiev.ua

Abstract: This study theoretically investigated light reflection and transmission in a system composed of a thin metal layer (Ag) adjacent to a rugate filter (RF) having a harmonic refractive index profile. Narrow dips in reflectance and peaks in transmittance in the RF band gap were obtained due to the excitation of a Tamm plasmon polariton (TPP) at the Ag–RF interface. It is shown that the spectral position and magnitude of the TPP dips/peaks in the RF band gap depend on the harmonic profile parameters of the RF refractive index, the metal layer thickness, and the external medium refractive index. The obtained dependences for reflectance and transmittance allow selecting parameters of the system which can be optimized for various applications.

Keywords: Tamm plasmon; Bragg mirror; rugate filter; band gap; light reflection and transmission

Citation: Reshetnyak, V.Y.; Pinkevych, I.P.; Bunning, T.J.; Evans, D.R. Influence of Rugate Filters on the Spectral Manifestation of Tamm Plasmon Polaritons. *Materials* **2021**, *14*, 1282. https://doi.org/10.3390/ma14051282

Academic Editor: Ivan V. Timofeev

Received: 13 February 2021
Accepted: 4 March 2021
Published: 8 March 2021

Publisher's Note: MDPI stays neutral with regard to jurisdictional claims in published maps and institutional affiliations.

1. Introduction

Recently, much attention has been paid to the study of Tamm plasmons and their applications. Tamm plasmon (otherwise known as Tamm plasmon polariton—TPP) is an electromagnetic mode localized at the interface between a metal film and a dielectric Bragg mirror [1–5]. Kaliteevsky et al. [3] called these localized modes Tamm plasmons by analogy with the electron states localized near the surface of a solid crystal and predicted by Tamm [6]. TPP localization is provided by the negative dielectric constant of the metal on the one side and the photonic stop band of the Bragg mirror on the other. In contrast to ordinary surface plasmon polaritons that are only transverse magnetic (TM) -polarized and have the dispersion relations outside the light cone, TPPs can be transverse electric (TE)- and TM-polarized with the dispersion inside the light cone. This allows direct optical excitation of TPP with TE- and TM-polarized light at any angle of incidence without the need for a prism or grating [3–7]. These unique capabilities make TPPs very attractive for various photoelectronic applications.

TPPs manifest themselves optically in the form of narrow dips/peaks in the reflection/transmission spectra in the spectral region corresponding to the photonic band gap of the Bragg mirror. TPPs are a good alternative to conventional surface plasmons, with potential applications for sensors [8–14], Tamm plasmon-based lasers [15–17], optical switches and filters [18–20], liquid crystal-tuned Tamm plasmon devices [21–23], and selective thermal and light emitters [24–28].

Over the past couple of decades, the optical properties of rugate filters (RFs) have been intensively studied [29]. RFs are dielectric thin films with a smooth periodic profile of the refractive index, giving rise to spectral band gaps like Bragg mirrors, typically composed of a square wave profile of the refractive index. The smooth profile of the RF refractive index makes it possible to improve many characteristics of optical devices compared to dielectric multilayer Bragg mirrors. RFs provide a photonic band gap without significant ripples in the reflection spectrum outside the band gap and without its higher harmonics, and they enable the possibility to overlay multiple harmonic waves (giving rise to multiple spectral notches). Furthermore, RFs have substantially higher laser-induced damage thresholds with respect to Bragg mirrors (see, for example, [29–32]). There have been numerous methods developed for obtaining RFs [33–37]; for example, porous silicon-based RFs represent a popular approach [31,38–42].

In this paper, RFs are explored as structures to excite Tamm plasmons, replacing the previously studied multilayer Bragg mirrors. The influence of RF parameters on the TPP excitation at the metal–RF interface is theoretically studied. The paper is organized as follows. Section 2 introduces a model of the metal–RF structure and derives equations allowing for the calculation of reflectance and transmittance of this structure in the RF band gap. Results of numerical calculations for a system using an Ag layer and their discussion are presented in Section 3. Section 4 presents some brief conclusions.

2. Theoretical Model and Basic Equations

Consider a structure, composed of an RF with a periodic dielectric function along the *z*-axis and a metal layer adjacent to the RF. A light beam, polarized along the *x*-axis, is normally incident on the metal layer along the *z*-axis and propagates through the metal and adjacent dielectric RF. Assuming that the principal axes of the RF dielectric tensor coincide with the Cartesian axes, the Cartesian indices of the electric and magnetic vectors can be omitted. A schematic of the structure together with directions of the light beams propagating in the system is presented in Figure 1.

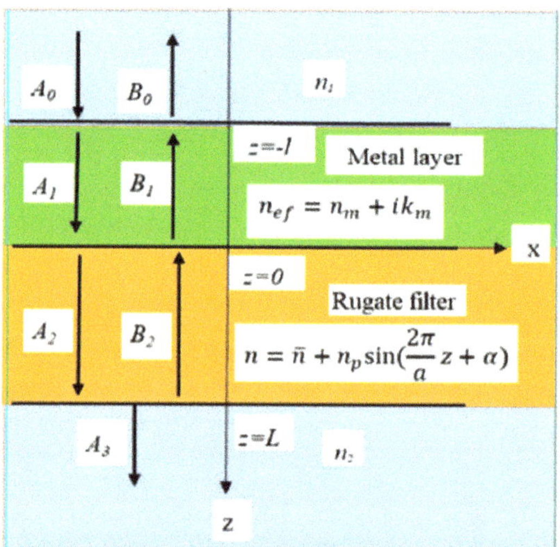

Figure 1. Schematic of the "metal layer–rugate filter (RF)" structure together with directions of the light beams propagating in the system and the refractive indices of the constituent substances.

In the area above the metal layer, $z \leq -l$, the electromagnetic field of the incident and reflected beams is described by the electric and magnetic vectors,

$$E_0(z) = A_0 \exp(ikn_1 z) + B_0 \exp(-ikn_1 z),$$
$$H_0(z) = \frac{n_1}{\mu_0 c}[A_0 \exp(ikn_1 z) - B_0 \exp(-ikn_1 z)] \tag{1}$$

where $k = \omega/c$, n_1 is a refractive index of the medium in front of the metal layer, and A_0, B_0 are the amplitudes of the incident and reflected beams, respectively.

In the metal layer, $-l \leq z \leq 0$, the electric and magnetic vectors of the wave field are written as

$$E_1(z) = A_1 \exp(ikn_{ef}z) + B_1 \exp(-ikn_{ef}z),$$
$$H_1(z) = \frac{n_{ef}}{\mu_0 c}[A_1 \exp(ikn_{ef}z) - B_1 \exp(-ikn_{ef}z)] \tag{2}$$

where $n_{ef} = n_m + ik_m$ is the complex refractive index of the metal, and A_1, B_1 are the amplitudes of the forward and backward waves, respectively.

In the RF area, $0 \leq z \leq L$, the dielectric tensor is periodic along the z-axis. Designating the period with the letter a, the dielectric tensor principal value $\varepsilon_{xx}(z)$ presented in the wave equation can be expanded in a Fourier series,

$$\varepsilon_{xx}(z) = \varepsilon_0 + \sum_{\substack{m=-\infty \\ (m \neq 0)}}^{\infty} \varepsilon_m e^{i\frac{2\pi}{a}mz} \tag{3}$$

For the sake of simplicity, neglecting absorption in the RF, $\varepsilon_0 = \bar{n}^2$ can be used in Equation (3), where $\bar{n}$ is the RF average refractive index. Next, consider the case when a wavelength of the incident beam is close to the Bragg wavelength λ_r satisfying the Bragg resonance condition $\frac{\lambda_r}{\bar{n}}m = 2a$, where m is an integer. Then, solving the wave equation in the RF area, one can use the coupled wave method [43,44] and present the electric vector of the electromagnetic field in the form of a superposition of the forward and backward waves,

$$E_2(z) = A_2(z) \exp(ik\bar{n}z) + B_2(z) \exp(-ik\bar{n}z) \tag{4}$$

where $A_2(z)$ and $B_2(z)$ are the slowly varying functions satisfying the Kogelnik Equations [39],

$$\frac{\partial A_2(z)}{\partial z} = i\chi_m \exp(-i2\delta z)B_2(z),$$
$$\frac{\partial B_2(z)}{\partial z} = -i\chi_{-m} \exp(i2\delta z)A_2(z) \tag{5}$$

In Equation (5) $\delta = k\bar{n} - \frac{\pi}{a}m$ is the offset from the Bragg resonance, $\chi_{\pm m} = k\varepsilon_{\pm m}/2\bar{n}$. Solving Equation (5) under the assumptions,

$$|\delta|a << 1, \ |\chi_{\pm m}|a << 1 \tag{6}$$

one can obtain solutions in the following form [45]:

$$A_2(z) = [a_1 \exp(-\gamma_m z) + a_2 \exp(\gamma_m z)] \exp(-i\delta z),$$
$$B_2(z) = \left[a_1 r_m \exp(-\gamma_m z) + a_2 \frac{\chi_{-m}}{\chi_m}\frac{1}{r_m} \exp(\gamma_m z)\right] \exp(i\delta z) \tag{7}$$

where $\gamma_m \approx \left(\chi_m\chi_{-m} - \delta^2\right)^{1/2}$, $r_m = \frac{i\chi_{-m}}{\gamma - i\delta}$, and coefficients a_1, a_2 can be determined from the boundary conditions.

Using the Maxwell equation $rot\vec{E} = -\partial\vec{B}/\partial t$ and Equations (4) and (5), the following expression is obtained for the magnetic field vector in the RF area:

$$H_2(z) = -\frac{1}{\mu_0 c}\left[\frac{\chi_{-m}}{k} \exp[-i(k\bar{n} - 2\delta)z] - \bar{n}\exp(ik\bar{n}z)\right]A_2(z) +$$
$$+\frac{1}{\mu_0 c}\left[\frac{\chi_m}{k} \exp[i(k\bar{n} - 2\delta)z] - \bar{n}\exp(-ik\bar{n}z)\right]B_2(z) \tag{8}$$

In the area below RF, $z \geq L$, there is only an outgoing wave described by the electric and magnetic vectors

$$E_3(z) = A_3 \exp(ikn_2 z), \ H_3(z) = \frac{n_2}{\mu_0 c}A_3 \exp(ikn_2 z) \tag{9}$$

where n_2 is a refractive index of the medium in the area $z \geq L$.

Now, to get expressions for the amplitudes of the reflected, B_0, and transmitted, A_3, waves, one can write down the boundary conditions for the electric and magnetic vectors at $z = -l$, $z = 0$ and $z = L$:

$$A_0 \exp(-ikn_1 l) + B_0 \exp(ikn_1 l) = A_1 \exp(-ikn_{ef}l) + B_1 \exp(ikn_{ef}l),$$
$$A_0 \exp(-ikn_1 l) - B_0 \exp(ikn_1 l) = \frac{n_{ef}}{n_1}[A_1 \exp(-ikn_{ef}l) - B_1 \exp(ikn_{ef}l)] \tag{10}$$

$$A_1 + B_1 = A_2(0) + B_2(0),$$
$$n_{ef}(A_1 - B_1) = -\left(\frac{\chi_{-m}}{k} - \overline{n}\right) A_2(0) + \left(\frac{\chi_m}{k} - \overline{n}\right) B_2(0) \tag{11}$$

$$A_2(L) \exp(ik\overline{n}L) + B_2(L) \exp(-ik\overline{n}L) = A_3 \exp(ikn_2 L),$$
$$\left(-\frac{\chi_{-m}}{k} + \overline{n}\right) A_2(L) \exp(i\delta L) + \left(\frac{\chi_m}{k} - \overline{n}\right) B_2(L) \exp(-i\delta L) = (-1)^m p^p n_2 A_3 \exp(ikn_2 L) \tag{12}$$

when writing Equation (12), the conditions $(k\overline{n} - \delta)L = m\pi L/a$ and $L/a = p$ are used, where p is the number of periods of dielectric function along the RF length.

Substituting Equation (7) into Equations (10)–(12), these equations can be solved, and expressions for the reflectance, $R = |B_0/A_0|^2$, and transmittance, $T = (n_2/n_1)|A_3/A_0|^2$, of the system can be obtained in the following form:

$$R = \left| 1 - \frac{\frac{2n_{ef}}{n_1}[1 - C + \exp(-i2kn_{ef}l)]}{(1 - \frac{n_{ef}}{n_1})(C - 1) + (1 + \frac{n_{ef}}{n_1})\exp(-i2kn_{ef}l)} \right|^2 \tag{13}$$

$$T = 16\frac{n_2}{n_1} \left| \frac{(1+r_m)\left(c_3 + \frac{\chi_{-m}}{\chi_m}\frac{1}{r_m}c_4\right) - (1 + \frac{\chi_{-m}}{\chi_m}\frac{1}{r_m})(c_3 + r_m c_4)}{(c_1 + r_m c_2)\left(c_3 + \frac{\chi_{-m}}{\chi_m}\frac{1}{r_m}c_4\right)\exp(\gamma_m L) - \left(c_1 + \frac{\chi_{-m}}{\chi_m}\frac{1}{r_m}c_2\right)(c_3 + r_m c_4)\exp(-\gamma_m L)} \right|^2 \times$$
$$\left| \frac{1}{(1 - \frac{n_{ef}}{n_1})(C-1)\exp(ikn_{ef}l) + (1 + \frac{n_{ef}}{n_1})\exp(-ikn_{ef}l)} \right|^2 \tag{14}$$

where

$$C = 2\frac{(1 + r_m)\left(c_3 + \frac{\chi_{-m}}{\chi_m}\frac{1}{r_m}c_4\right)\exp(\gamma_m L) - (1 + \frac{\chi_{-m}}{\chi_m}\frac{1}{r_m})(c_3 + r_m c_4)\exp(-\gamma_m L)}{(c_1 + r_m c_2)\left(c_3 + \frac{\chi_{-m}}{\chi_m}\frac{1}{r_m}c_4\right)\exp(\gamma_m L) - \left(c_1 + \frac{\chi_{-m}}{\chi_m}\frac{1}{r_m}c_2\right)(c_3 + r_m c_4)\exp(-\gamma_m L)} \tag{15}$$

$$c_1 = 1 + \frac{\overline{n}}{n_{ef}} - \frac{\chi_{-m}}{k}\frac{1}{n_{ef}}, \quad c_2 = 1 - \frac{\overline{n}}{n_{ef}} + \frac{\chi_m}{k}\frac{1}{n_{ef}},$$
$$c_3 = (-1)^{mp}\left(\overline{n} - n_2 - \frac{\chi_{-m}}{k}\right), \quad c_4 = (-1)^{mp}\left(-\overline{n} - n_2 + \frac{\chi_m}{k}\right) \tag{16}$$

To consider the case when the RF refractive index is a periodic function, the following form is used:

$$n(z) = \overline{n} + n_p \sin\left(\frac{2\pi}{a}z + \alpha\right) \tag{17}$$

In Equation (17), varying α changes the value of the refractive index in the locations immediately adjacent to the metal; this gives a more general form of the harmonic function of the refractive index. Assuming $n_p \ll \overline{n}$, the corresponding dielectric function is $\varepsilon_{xx}(z) \approx \overline{n}^2 + 2n_p\overline{n}\sin\left(\frac{2\pi}{a}z + \alpha\right)$. According to Equation (3), it has the following non-zero Fourier components: $\varepsilon_0 = \overline{n}^2$, $\varepsilon_1 = -in_p\overline{n}e^{i\alpha}$, $\varepsilon_{-1} = in_p\overline{n}e^{-i\alpha}$. In this case, in Equations (13)–(16), the integer m must be set to 1; therefore, the Bragg wavelength $\lambda_r = 2\overline{n}a$.

3. Results of Numerical Calculations and Discussion

For numerical calculations, we take the "Ag layer–RF" structure with parameters that are close to those previously studied for a system consisting of an Ag plate and a Bragg mirror composed of alternating TiO$_2$ and SiO$_2$ layers with a thickness of $d_1 = 50.4$ nm and $d_2 = 86.7$ nm, respectively. In this system, the Tamm plasmon resonances were experimentally detected in the visible region [46]; therefore, we take the same period for the RF refractive index, $a = d_1 + d_2 = 137.1$ nm, and average refractive index $\bar{n} = 1.85$, calculated as $\bar{n} = \left(n_{TiO_2}d_1 + n_{SiO_2}d_2\right)/(d_1 + d_2)$ where n_{TiO_2} and n_{SiO_2} are the refractive indices of TiO$_2$ and SiO$_2$, respectively [47,48].

The conditions in Equation (6) impose restrictions on the maximum values of a magnitude n_p of the RF refractive index modulation. Indeed, the RF band gap is proportional to $\chi_{\pm 1}$ [45] and, therefore, to n_p (taking into account that $|\chi_{\pm 1}| = |k\varepsilon_{\pm 1}/2\bar{n}| \approx \pi n_p/\lambda_r$). The value $n_p = 0.2$ or values close to it ensure the fulfillment of the conditions in Equation (6) and are used further in the numerical calculations. For the complex refractive index of Ag, frequency dispersion is taken into account [49]. The refractive indices of media before the Ag layer and after the RF film are parameters that can be varied.

We calculated the reflectance and transmittance spectra of the system composed of the RF film with a refractive index profile described by Equation (17) and the Ag layer placed at the top of the RF (see Figure 1). For calculations, we set the refractive indices of the media before the Ag layer and after the RF as $n_1 = n_2 = 1$, the Ag layer thickness as 45 nm, and the RF thickness as $L = 1919.4$ nm (i.e., 14 periods of the RF refractive index). Values of α in Equation (17) used for the calculations are presented in Table 1 and selected only as an example.

Table 1. The RF refractive indices used for numerical calculations.

α	Rugate Filter Refractive Index	Spectral Band Number in Figure 2
$-\pi/2$	$n(z) = \bar{n} - n_p\cos(2\pi z/a)$	1
0	$n(z) = \bar{n} + n_p\sin(2\pi z/a)$	2
$\pi/2$	$n(z) = \bar{n} + n_p\cos(2\pi z/a)$	3
π	$n(z) = \bar{n} - n_p\sin(2\pi z/a)$	4

For each initial (at $z = 0$) phase α of the RF refractive index, reflection dips and transmission peaks in the spectral region of the RF band gap were obtained, associated with the excitation of TPP at the Ag layer–RF interface. Results of calculations are shown in Figure 2a for all α values presented in Table 1. Figure 2b shows the reflectance of only the RF with the same parameters that were used for each case of α in Figure 2a. As can be seen, the inclusion of the thin layer of metal on top of the RF thin film drastically modified the spectral content of the optical architecture.

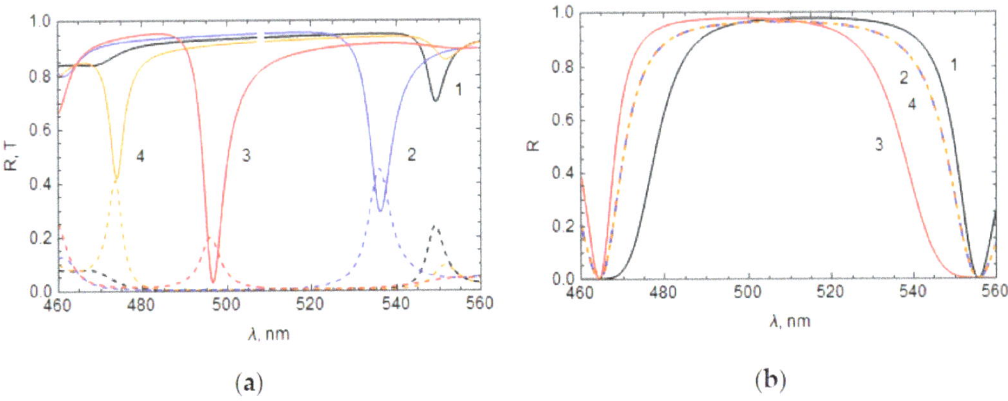

(a) **(b)**

Figure 2. (**a**) Reflectance and transmittance spectra of the "Ag layer–RF" system at different initial phase α: $\alpha = -\pi/2$ (1, black), 0 (2, blue), $\pi/2$ (3, red), and π (4, orange); reflectance—Solid lines, transmittance—Dashed lines, Ag film thickness = 45 nm. (**b**) Reflectance spectrum of only the RF, where the numbers near the curves correspond with the α values used in Figure 2a.

It can be seen from Figure 2a that the position and magnitude of the TPP spectral bands depend strongly on the initial phase α, as the TPP wavelength decreases with increasing α (Figure 2b). Comparing the RF refractive index profiles, which correspond to bands 1 and 3 $[n(z) = \bar{n} \pm n_p \cos(2\pi z/a)]$ or 2 and 4 $n(z) = \bar{n} \pm n_p \sin(2\pi z/a)$ we can conclude that the sign of the term added to the RF average refractive index, $\bar{n}$, significantly affects the TPP wavelength.

The influence of the Ag layer thickness on the TPP wavelength and the TPP dip/peak magnitude is shown in Figure 3 for the bands 1 (Figure 3a), 2 (Figure 3b), 3 (Figure 3c), and 4 (Figure 3d). The TPP wavelength and reflectance dip, as a rule, decrease (the TPP transmittance peak increases) with an increase in Ag layer thickness; however, for peak 3, there is an optimal thickness of 35 nm.

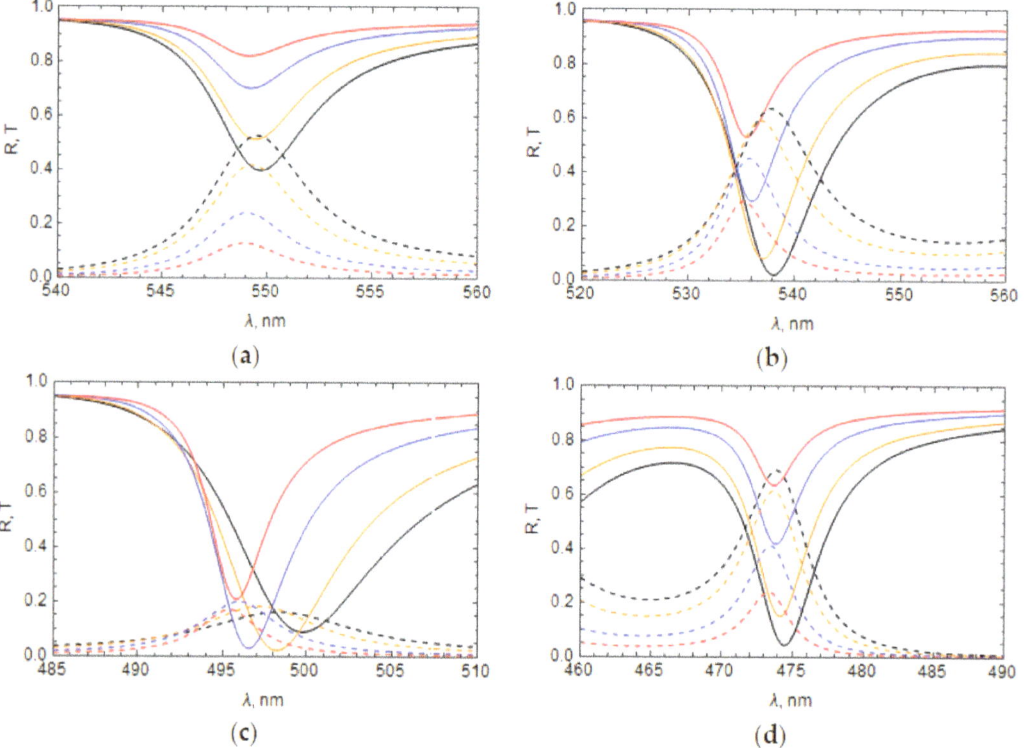

Figure 3. Influence of the Ag film thickness l on the reflectance and transmittance of the system for different RF profiles $n(z)$; (**a**) TPP band 1, $n(z) = \bar{n} - n_p \cos(2\pi z/a)$; (**b**) TPP band 2, $n(z) = \bar{n} + n_p \sin(2\pi z/a)$; (**c**) TPP band 3, $n(z) = \bar{n} + n_p \cos(2\pi z/a)$; (**d**) TPP band 4, $n(z) = \bar{n} - n_p \sin(2\pi z/a)$. Reflectance—Solid lines, transmittance—Dashed lines, l = 30 nm—Black, 35 nm—Orange, 45 nm—Blue, 55 nm—Red. The TPP bands refer to those found in Figure 2.

In Figure 4, we show the influence of the refractive index of the medium above the Ag layer, n_1, on the reflectance and transmittance of the system with the RF profiles $n(z)$ from Table 1. In all of these cases, an increase in the refractive index of the medium above the Ag layer leads to a TPP wavelength decrease and a change in the dip/peak magnitude.

The influence of the refractive index of the medium below the RF, n_2, on the reflectance and transmittance of the system is shown in Figure 5 for the same cases as in Figure 4. As in the case of a change of the refractive index of the medium above the Ag layer, an increase in the medium refractive index below the RF shifts the TPP bands toward shorter wavelengths and changes the dip/peak magnitude.

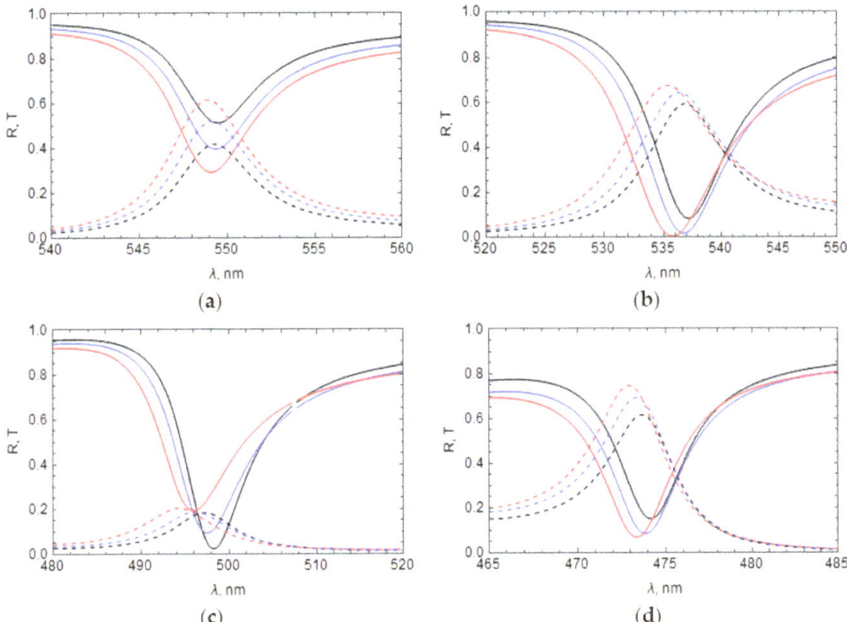

Figure 4. Influence of the refractive index of the medium above the Ag layer, n_1, on the reflectance and transmittance: $n_1 = 1$ (black line), 1.5 (blue), and 2.5 (red); (**a**) TPP band 1; (**b**) TPP band 2; (**c**) TPP band 3; (**d**) TPP band 4. Reflectance—Solid lines, transmittance—Dashed lines, thickness of the Ag layer = 35 nm.

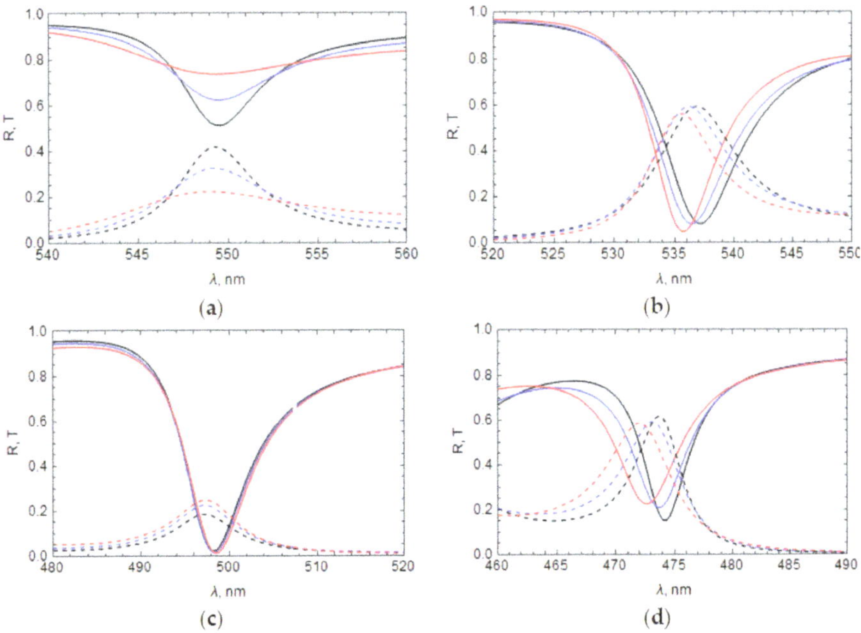

Figure 5. Influence of the refractive index of medium below the RF, n_2 on reflectance and transmittance: $n_2 = 1$ (black line), 1.5 (blue line), and 2.5 (red line); (**a**) TPP band 1; (**b**) TPP band 2; (**c**) TPP band 3; (**d**) TPP band 4. Reflectance—Solid lines, transmittance—Dashed lines, thickness of the Ag layer = 35 nm.

As seen from Figures 4 and 5, the dependence of reflectance and transmittance on the refractive index of the media above the metal layer and below the RF is different for different profiles $n(z)$ of the RF. Furthermore, the impact of n_2 becomes weaker when the transmittance of the system decreases. It takes place for some profiles $n(z)$ (see, for example, Figure 4c) or with increasing the number of periods $n(z)$ along the RF. In both of these cases, the electromagnetic field at the boundary of the RF with the adjacent medium (after the RF) becomes weaker; therefore, the influence of the adjacent medium on TPP is also weakened.

The TPP wavelength is proportional to the Bragg wavelength $\lambda_r = 2a\bar{n}$ [3], and there is an obvious shift in the spectral position of the TPP bands with a change in the RF period a and the average refractive index $\bar{n}$. However, the influence of the magnitude of the RF refractive index modulation, n_p, is not obvious. In Figure 6, we show reflectance and transmittance of the system with fixed a and $\bar{n}$, but with different n_p values for the RF refractive index profiles $n(z)$ taken from Table 1.

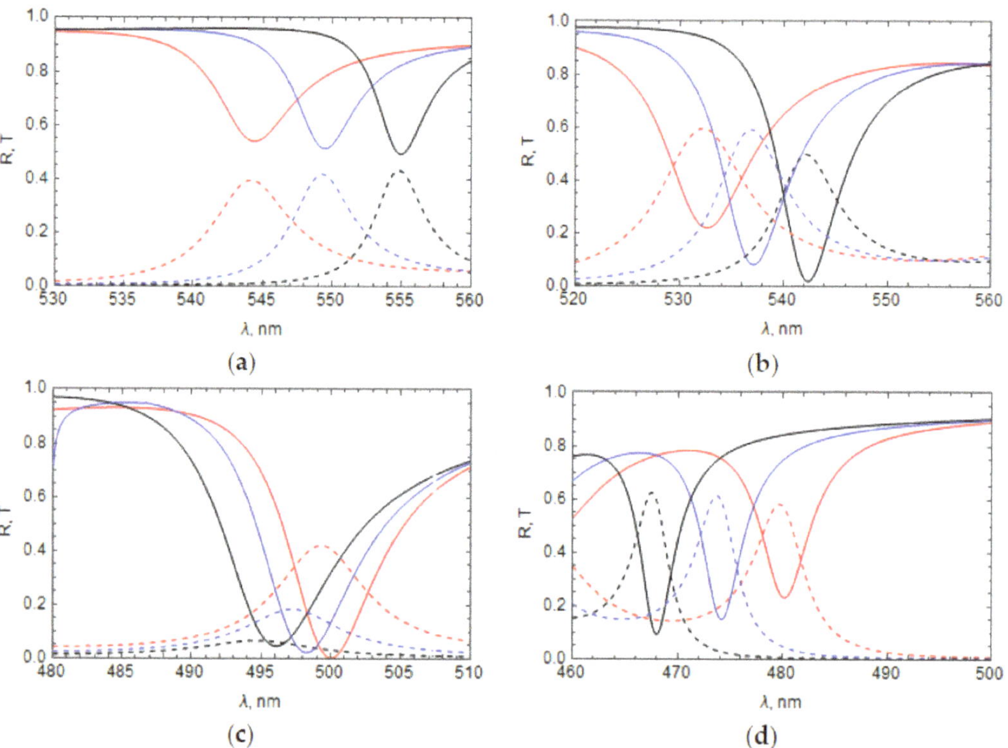

Figure 6. Reflectance and transmittance of the system at different values n_p = 0.15 (red), 0.2 (blue), and 0.25 (black); (**a**) TPP band 1; (**b**) TPP band 2; (**c**) TPP band 3; (**d**) TPP band 4. Thickness of the Ag layer = 35 nm, a = 137.1 nm, $\bar{n} = 1.85$.

The RF band gap is proportional to n_p and, therefore, broadens with increasing n_p. This broadening leads to a TPP wavelength shift (see Figure 6), which has an opposite sign for the peaks to the right and left of the RF band gap center (see Figure 2a,b). As a result, the TPP wavelength in the case of bands 1 and 2 increases with increasing n_p (Figure 6a,b, respectively), while the TPP wavelength in the case of the bands 3 and 4 decreases with increasing n_p (Figure 6c,d, respectively). The magnitude of the reflectance dips and transmittance peaks can change with increasing n_p depending on the dip/peak spectral position.

Calculations were also performed for the case when the Ag layer is placed at the bottom of RF. The character of the dependence of reflectance and transmittance on the parameters of the system in this case is the same as in the case of the Ag layer on the top of RF. However, the degree of influence of these parameters decreases significantly with an increasing number of RF periods due to a strong decrease in the electromagnetic field exciting the TPP at the bottom of the RF.

We also compared the results of calculating the reflectance and transmission spectra at $n_p = 0.2$ using Equations (13)–(16) to results obtained using COMSOL modeling. A difference between these cases is barely noticeable.

Lastly, we also calculated the reflectance spectrum of the "Ag layer + RF" system using Equation (13), slightly going beyond the conditions in Equation (6) for applicability of the obtained analytical solution. For this, we set the magnitude of the RF refractive index modulation to be $n_p = 0.5$. In this case, $|\chi_{\pm 1} a| \approx 0.43$ and the second condition in Equation (6) is violated. Results of the calculation for all cases of the RF refractive index profile $n(z)$ presented in Table 1 at $n_p = 0.5$ are shown in Figure 7 (dotted curves). In the same figure, we also show the results of calculating the reflectance spectrum obtained using COMSOL software (solid curves).

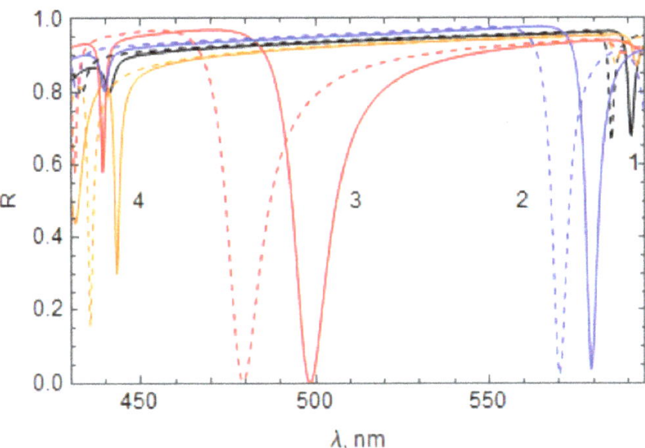

Figure 7. Comparison of the reflectance spectrum of the "Ag layer + RF" calculated using COMSOL software (solid lines) and using Equations (12)–(15) (dashed lines) at $n_p = 0.5$ TPP bands are numbered according to Table 1: 1 (black), 2 (blue), 3 (red), 4 (orange). $\bar{n} = 1.85$, $a = 137.1$ nm, thickness of the Ag layer = 45 nm.

It can be seen that, for all considered cases of $n(z)$, the qualitative picture of the spectral distribution of TPP bands obtained using the analytical solution remains the same as in the calculation using COMSOL software. This holds true even when there is a slight violation of the condition in Equation (6), although, in this case, the wavelength of the TPP bands is slightly shifted toward shorter waves, as shown in Figure 7.

4. Conclusions

We studied the light reflectance and transmittance of a system composed of a metal (Ag) layer adjacent to a rugate filter having a harmonic refractive index profile. Narrow dips in the reflectance and peaks in the transmittance were obtained, due to the excitation of TPP at the Ag layer–RF interface. We show that parameters of the harmonic profile of the RF refractive index significantly affect the TPP wavelength and magnitude of the TPP dips/peaks. Depending on the profile of the RF refractive index, the spectral position of the TPP can be at any point in the RF band gap. The influence of the metal layer thickness and the external medium refractive index on the position and magnitude of the TPP dips/peaks was also established. It should be noted that the proposed analytical solution describes the spectral position and magnitude of the TPP bands quite well, even when the RF parameters slightly violate the conditions for its derivation.

Lastly, we would like to point out that the potential applications of TPPs on metal with adjacent RF include all areas where TPPs on metal with an adjacent traditional Bragg mirror are applicable, but with the advantages of RF mentioned in Section 1. In particular, the possibility of superimposing several harmonic waves of the refractive index in RF allows for the simultaneous use of several spectral notches, which makes it possible to use TPPs excited in the spectral region of different notches.

We believe that the proposed analytical method for studying plasmonic structures with an RF and the obtained dependences of reflectance and transmittance can be used for designing devices based on Tamm plasmons.

Author Contributions: Conceptualization and methodology, V.Y.R.; investigation and writing—original draft preparation, I.P.P.; conceptualization, validation, and writing—review and editing, T.J.B.; writing—review and editing and supervision, D.R.E. All authors have read and agreed to the published version of the manuscript.

Funding: This research was partially funded by STCU, grant number P652c, and the Ministry of Education and Science of Ukraine, grant number 20BF051-04.

Institutional Review Board Statement: Not applicable.

Informed Consent Statement: Not applicable.

Data Availability Statement: Data sharing not applicable.

Conflicts of Interest: The authors declare no conflict of interest. The funders had no role in the design of the study; in the collection, analyses, or interpretation of data; in the writing of the manuscript, or in the decision to publish the results.

References

1. Gaspar-Armenta, A.; Villa, F. Photonic surface-wave excitation: Photonic crystal-metal interface. *J. Opt. Soc. Am. B* **2003**, *20*, 2349–2354. [CrossRef]
2. Vinogradov, A.P.; Dorofeenko, A.V.; Erokhin, S.G.; Inoue, M.; Lisyansky, A.A.; Merzlikin, A.M.; Granovsky, A.B. Surface state peculiarities in one-dimensional photonic crystal interfaces. *Phys. Rev. B* **2006**, *74*, 045128. [CrossRef]
3. Kaliteevski, M.; Iorsh, I.; Brand, S.; Abram, R.A.; Chamberlain, J.M.; Kavokin, A.V.; Shelykh, I.A. Tamm plasmon-polaritons: Possible electromagnetic states at the interface of a metal and a dielectric Bragg mirror. *Phys. Rev. B* **2007**, *76*, 165415. [CrossRef]
4. Sasin, M.E.; Seisyan, R.P.; Kalitteevski, M.A.; Brand, S.; Abram, R.A.; Chamberlain, J.M.; Egorov, A.Y.u.; Vasil'ev, A.P.; Mikhrin, V.S.; Kavokin, A.V. Tamm plasmon polaritons: Slow and spatially compact light. *Appl. Phys. Lett.* **2008**, *92*, 251112. [CrossRef]
5. Sasin, M.E.; Seisyan, R.P.; Kalitteevski, M.A.; Brand, S.; Abram, R.A.; Chamberlain, J.M.; Iorsh, I.; Shelykh, I.; Egorov, A.Y.U.; Vasil'ev, A.P.; et al. Tamm plasmon-polaritons: First experimental observation. *Superlattices Microstruct.* **2010**, *47*, 44–49. [CrossRef]
6. Tamm, I. Über eine mögliche Art der Elektronenbindung an Kristalloberflächen. *Z. Phys.* **1932**, *76*, 849–850. [CrossRef]
7. Takayama, O.; Bogdanov, A.A.; Lavrinenko, A.B. Photonic surface waves on metamaterial interfaces. *J. Phys. Cond. Mat.* **2017**, *29*, 463001. [CrossRef] [PubMed]
8. Zhang, W.L.; Wang, F.; Rao, Y.J.; Jiang, Y. Novel sensing concept based on optical Tamm plasmon. *Opt. Exp.* **2014**, *22*, 14524–14529. [CrossRef] [PubMed]
9. Auguié, B.; Fuertes, M.C.; Angelomé, P.C.; Abdala, N.L.; Soler Illia, G.J.A.A.; Fainstein, A. Tamm plasmon resonance in mesoporous multilayers: Toward a sensing application. *ACS Photonics* **2014**, *1*, 775–780. [CrossRef]
10. Baryshev, A.V.; Merzlikin, A.M. Approach to visualization of and optical sensing by Bloch surface waves in noble or base metal-based plasmonic photonic crystal slabs. *Appl. Opt.* **2014**, *53*, 3142–3146. [CrossRef] [PubMed]
11. Kumar, S.; Maji, P.S.; Das, R. Tamm-plasmon resonance based temperature sensor in a Ta_2O_5/SiO_2 based distributed Bragg reflector. *Sens. Actuators A* **2017**, *260*, 10–15. [CrossRef]
12. Maji, P.S.; Shukla, M.K.; Das, R. Blood component detection based on miniaturized self-referenced hybrid Tamm-plasmon-polariton sensor. *Sens. Actuators B* **2018**, *255*, 729–734. [CrossRef]
13. Buzavaite-Verteliene, E.; Plikusiene, I.; Tolenis, T.; Valavicius, A.; Anulyte, J.; Ramanavicius, A.; Balevicius, Z. Hybrid Tamm-surface plasmon polariton mode for highly sensitive detection of protein interactions. *Opt. Express* **2020**, *28*, 29033–29043. [CrossRef] [PubMed]
14. Balevicius, Z. Strong coupling between Tamm and surface plasmons for advanced optical bio-sensing. *Coatings* **2020**, *10*, 1187. [CrossRef]

15. Symonds, C.; Lemaître, A.; Senellart, P.; Jomaa, M.H.; Aberra Guebrou, S.; Homeyer, E.; Brucoli, G.; Bellessa, J. Lasing in a hybrid GaAs/silver Tamm structure. *Appl. Phys. Lett.* **2012**, *100*, 121122. [CrossRef]

16. Brückner, R.; Zakhidov, A.A.; Scholz, R.; Sudzius, M.; Hintschich, S.I.; Fröb, H.; Lyssenko, V.G.; Leo, K. Phase-locked coherent modes in a patterned metal–organic microcavity. *Nat. Photonics* **2012**, *6*, 322–326. [CrossRef]

17. Symonds, C.; Lheureux, G.; Hugonin, J.P.; Greffet, J.J.; Laverdant, J.; Brucoli, G.; Lemaitre, A.; Senellart, P.; Bellessa, J. Confined Tamm plasmon lasers. *Nano Lett.* **2013**, *13*, 3179–3184. [CrossRef]

18. Zhang, W.L.; Yu, S.F. Bistable switching using an optical Tamm cavity with a Kerr medium. *Opt. Commun.* **2010**, *283*, 2622–2626. [CrossRef]

19. Zhou, H.; Yang, G.; Wang, K.; Long, H.; Lu, P. Multiple optical Tamm states at a metal−dielectric mirror interface. *Opt. Lett.* **2010**, *35*, 4112–4114. [CrossRef]

20. Gong, Y.; Liu, X.; Lu, H.; Wang, I.; Wang, G. Perfect absorber supported by optical Tamm states in plasmonic waveguide. *Opt. Exp.* **2011**, *19*, 18393–18398. [CrossRef]

21. Cheng, H.-C.; Kuo, C.-Y.; Hung, Y.-J.; Chen, K.-P.; Jeng, S.-C. Liquid-Crystal Active Tamm-Plasmon Devices. *Phys. Rev. Appl.* **2018**, *9*, 064034. [CrossRef]

22. Timofeev, I.V.; Pankin, P.S.; Vetrov, S.Y.; Arkhipkin, V.G.; Lee, W.; Zyryanov, V.Y. Chiral Optical Tamm States: Temporal Coupled-Mode Theory. *Crystals* **2017**, *7*, 113. [CrossRef]

23. Buchnev, O.; Belosludtsev, A.; Reshetnyak, V.; Evans, D.R.; Fedotov, V.A. Observing and controlling a Tamm plasmon at the interface with a metasurface. *Nanophotonics* **2020**, *9*, 897–903. [CrossRef]

24. Yang, Z.-Y.; Ishii, S.; Yokoyama, T.; Dao, T.D.; Sun, M.-G.; Nagao, T.; Chen, K.-P. Tamm plasmon selective thermal emitters. *Opt. Lett.* **2016**, *41*, 4453–4456. [CrossRef] [PubMed]

25. Yang, Z.-Y.; Ishii, S.; Yokoyama, T.; Dao, T.D.; Sun, M.-G.; Pankin, P.S.; Timofeev, I.V.; Nagao, T.; Chen, K.-P. Narrowband wavelength selective thermal emitters by confined Tamm plasmon polaritons. *ACS Photonics* **2017**, *4*, 2212–2219. [CrossRef]

26. Lee, B.J.; Fu, C.J.; Zhang, Z.M. Coherent thermal emission from one-dimensional photonic crystals. *Appl. Phys. Lett.* **2005**, *87*, 071904. [CrossRef]

27. Gazzano, O.; Vasconcellos, S.M.; Gauthron, K.; Symonds, C.; Voisin, P.; Bellessa, J.; Lemaître, A.; Senellart, P. Single photon source using confined Tamm plasmon modes. *Appl. Phys. Lett.* **2012**, *100*, 232111. [CrossRef]

28. Jiménez-Solano, A.; Galisteo-López, J.F.; Míguez, H. Flexible and adaptable light-emitting coatings for arbitrary metal surfaces based on optical Tamm mode coupling. *Adv. Opt. Mater.* **2018**, *6*, 1700560. [CrossRef]

29. Bovard, B.G. Rugate filter theory: An overview. *Appl. Opt.* **1993**, *32*, 5427–5442. [CrossRef]

30. Southwell, W.H.; Hall, R.L. Rugate filter sidelobe suppression using quintic and rugated quintic matching layers. *Appl. Opt.* **1989**, *28*, 2949–2951. [CrossRef]

31. Lorenzo, E.; Oton, C.J.; Capuj, N.E.; Ghulinyan, M.; Navarro-Urrios, D.; Gaburro, Z.; Pavesi, L. Porous silicon-based rugate filters. *Appl. Opt.* **2005**, *44*, 5415–5421. [CrossRef] [PubMed]

32. Jupé, M.; Lappschies, M.; Jensen, L.; Starke, K.; Ristau, D. Laser-induced damage in gradual index layers and Rugate filters. *Proc. SPIE* **2006**, *6403*, 640311.

33. Bartholomew, C.S.; Morrow, M.D.; Betz, H.T.; Grieser, J.L.; Spence, R.A.; Murarka, N.P. Rugate filters by laser flash evaporation of SiOxNy on room-temperature polycarbonate. *J. Vac. Sci. Technol. A* **1988**, *6*, 1703–1707. [CrossRef]

34. Gunning, W.J.; Hall, R.L.; Woodberry, F.J.; Southwell, W.H.; Gluck, N.S. Codeposition of continuous composition rugate filters. *Appl. Opt.* **1989**, *28*, 2945–2948. [CrossRef] [PubMed]

35. Jankowski, A.F.; Schrawyer, L.R.; Perry, P.L. Reactive sputtering of molybdenum-oxide gradient-index filters. *J. Vac. Sci. Technol. A* **1991**, *9*, 1184–1187. [CrossRef]

36. Swart, P.L.; Bulkin, P.V.; Lacquet, B.M. Rugate filter manufacturing by electron cyclotron resonance plasma-enhanced chemical vapor deposition of SiNx. *Opt. Eng.* **1997**, *36*, 1214–1219. [CrossRef]

37. Kaminska, K.; Brown, T.; Beydaghyan, G.; Robbie, K. Rugate filters grown by glancing angle deposition. In *Applications of Photonic Technology 5*; Lessard, R.A., Lampropoulos, G.A., Schini, G.W., Eds.; SPIE: Bellingham, WA, USA, 2003; Volume 4833, pp. 633–639.

38. Berger, M.G.; Arens-Fischer, R.; Thönissen, M.; Krüger, M.; Billat, S.; Lüth, H.; Hilbrich, S.; Theiss, W.; Grosse, P. Dielectric filters made of PS: Advanced performance by oxidation and new layer structures. *Thin Solid Films* **1997**, *297*, 237–240. [CrossRef]

39. Kaminska, K.; Brown, T.; Beydaghyan, G.; Robbie, K. Vacuum evaporated porous silicon photonic interference filters. *Appl. Opt.* **2003**, *42*, 4212–4219. [CrossRef]

40. Keshavarzi, S.; Kovacs, A.; Abdo, M.; Badilita, V.; Zhu, R.; Korvink, J.G.; Mescheder, U. Porous silicon based rugate filter wheel for multispectral imaging applications. *ECS J. Sol. St. Sci. Tech.* **2019**, *8*, Q43–Q49. [CrossRef]

41. Ilyasa, S.; Böckinga, T.; Kilianb, K.; Reecea, P.J.; Goodingb, J.; Gausc, K.; Gala, M. Porous silicon based narrow line-width rugate filters. *Opt. Mater.* **2007**, *29*, 619–622. [CrossRef]

42. Verly, P.G. Hybrid approach for rugate filter design. *Appl. Opt.* **2008**, *47*, C172–C178. [CrossRef] [PubMed]

43. Kogelnik, H. Coupled wave theory for thick hologram gratings. *Bell Syst. Tech. J.* **1969**, *48*, 2909–2947. [CrossRef]

44. Yariv, A.; Yeh, P. *Optical Waves in Crystals: Propagation and Control of Laser Radiation*; J. Wiley & Sons: Hoboken, NJ, USA, 2003; pp. 177–201.

45. Karpov, S.Y.; Stolyarov, S.N. Propagation and transformation of electromagnetic waves in one-dimensional periodic structures. *Phys. Usp.* **1993**, *36*, 1–22. [CrossRef]

46. Chang, C.-Y.; Chen, Y.-H.; Tsai, Y.-L.; Kuo, H.-C.; Chen, K.-P. Tunability and optimization of coupling efficiency in Tamm plasmon modes. *IEEE J. Sel. Top. Quant. Electr.* **2015**, *21*, 4600206. [CrossRef]
47. Siefke, T.; Kroker, S.; Pfeiffer, K.; Puffky, O.; Dietrich, K.; Franta, D.; Ohlídal, I.; Szeghalmi, A.; Kley, E.-B.; Tünnermann, A. Materials pushing the application limits of wire grid polarizers further into the deep ultraviolet spectral range. *Adv. Opt. Mater.* **2016**, *4*, 1780–1786. [CrossRef]
48. Gao, L.; Lemarchand, F.; Lequime, M. Refractive index determination of SiO_2 layer in the UV/Vis/NIR range: Spectrophotometric reverse engineering on single and bi-layer designs. *J. Eur. Opt. Soc. Rap. Publ.* **2013**, *8*, 13010. [CrossRef]
49. Rakić, A.D.; Djurišic, A.B.; Elazar, J.M.; Majewski, M.L. Optical properties of metallic films for vertical-cavity optoelectronic devices. *Appl. Opt.* **1998**, *37*, 5271–5283. [CrossRef] [PubMed]

Article

Chiral Optical Tamm States at the Interface between a Dye-Doped Cholesteric Liquid Crystal and an Anisotropic Mirror

Anastasia Yu. Avdeeva [1,*], Stepan Ya. Vetrov [1,2], Rashid G. Bikbaev [1,2], Maxim V. Pyatnov [1,2], Natalya V. Rudakova [1,2] and Ivan V. Timofeev [1,2]

[1] Kirensky Institute of Physics, Federal Research Center KSC SB RAS, Krasnoyarsk 660036, Russia;
S.Vetrov@inbox.ru (S.Y.V.); bikbaev@iph.krasn.ru (R.G.B.); MaksPyatnov@yandex.ru (M.V.P.);
Natalya-V-Rudakova@iph.krasn.ru (N.V.R.); tiv@iph.krasn.ru (I.V.T.)
[2] Siberian Federal University, Krasnoyarsk 660041, Russia
* Correspondence: Anastasia-yu-avdeeva@iph.krasn.ru

Received: 2 July 2020; Accepted: 21 July 2020; Published: 22 July 2020

Abstract: The resonant splitting of optical Tamm state numerically is demonstrated. The Tamm state is localized at the interface between a resonant chiral medium and a polarization-preserving anisotropic mirror. The chiral medium is considered as a cholesteric liquid crystal doped with resonant dye molecules. The article shows that the splitting occurs when dye resonance frequency coincides with the frequency of the Tamm state. In this case the reflectance, transmittance, and absorptance spectra show two distinct Tamm modes. For both modes, the field localization is at the interface between the media. The external field control of configurable optical and structural parameters paves the way for use in tunable chiral microlaser.

Keywords: localization of light; photonic crystals; chirality; dye-doped cholesteric liquid crystal; optical Tamm states; resonant frequency dispersion

---

1. Introduction

Recently, there have been an increasing number of fundamental and applied works devoted to searching for new promising materials and designing the structures that exploit new ways of controlling light. Of particular interest is the optical Tamm state (OTS), i.e., a surface state localized at the interface between two media serving as mirrors, which does not transfer energy along the interface and exponentially decreases with increasing distance on either side of the interface [1–3]. The OTS is an electromagnetic analog of the Tamm state of electrons at the superlattice boundary [4]. In experiments, the OTS manifests itself as a narrow peak in the energy spectra of a sample [4,5]. The interest in the OTSs is due to the potential of their application in lasers and emitters [6–8], absorbers [9,10], sensors [11,12], as well as in photovoltaics [13], topological photonics [14,15], and other devices [16–19].

It appeared a nontrivial task to induce the OTS at the interface between an isotropic medium and a chiral medium, such as a cholesteric liquid crystal (CLC). A CLC is formed by oriented elongated molecules with the preferred direction twisted in space in the form of a helix. The CLCs are characterized by the continuous helical symmetry of the permittivity tensor and, due to its periodicity, represent one-dimensional photonic crystals [20]. The CLCs attract attention by their high sensitivity to electric and magnetic fields and temperature variation [21]. The qualitative difference between the CLCs and other types of photonic-crystal structures is that the former exhibit the diffraction selectivity to the polarization of light. Therefore, to localize a surface state at the interface between a

CLC and an isotropic mirror, which does not preserve the polarization of light upon reflection, it is necessary to compensate the polarization variation using a quarter-wave plate [22], an additional anisotropic layer [23], or a chirality-preserving mirror [24]. Such mirror can preserve not only the chirality sign, but also the value of the ellipticity of the incident radiation. This particular case can be called a polarization-preserving anisotropic mirror (PPAM) or a mirror that does not change the sign of polarization of the reflected light [25]. The simplest example of such structures is a pile of identical anisotropic layers with alternating orientations of the optical axes. This structure was studied by Reusch [26] as a polarization filter. Due to the chiral properties of the Reusch pile, the incident wave of one circular polarization passes through the structure, while the component of the other polarization is reflected. This property is observed both at the normal [27] and oblique incidence of light [28]. A special class is the equichiral Reusch piles, in which the optical axes of neighboring layers become perpendicular to each other [29]. Previously, we demonstrated the possibility of implementing the OTS at the interface between a PPAM and a CLC; the obtained localized surface state was called the chiral optical Tamm state (COTS) [23,24,30,31].

Liquid crystals doped with various micro- and nanoparticles [32] or dye molecules evoke great interest, since they combine the fluidity, crystal anisotropy, and specific properties of dye particles or molecules. In a dye-doped CLC, a distributed feedback lasing with the lowest laser pumping threshold can be implemented [33–36]. The presence of dye molecules can lead to the qualitative rearrangement of the band structure of the CLC spectrum, specifically, to splitting of the photonic band gap (PBG) into several PBGs [37]. Embedding of a resonant defect layer doped with metal nanoparticles into a CLC adds new features to the spectral and polarization properties of the latter [38].

In view of the persistent popularity of the discussed topic, we set a problem to examine the spectral properties of the COTS localized at the interface between a PPAM and a dye-doped cholesteric liquid crystal (DDCLC). The advantage of the proposed structure is the significant expansion of the possibility of effective controlling the parameters of the photon energy spectrum and the transmittance, reflectance, and absorptance spectra of the structure. A fundamentally new effect of COTS splitting at the interface between the media at the coinciding dye and COTS resonant frequencies was established. In this case, two energy levels corresponding to two new COTSs were observed at the intersection of the DDCLC and PPAM band gaps. The complete or partial overlap of the band gaps of CLC and PPAM in the energy spectrum is a condition for the COTS formation at the two environments border plane. In this case, the energy level occurs at the overlap of the band gaps corresponding to the chiral optical Tamm mode localized at the interface of the environments. The dye molecules included in the CLC are influenced by an electromagnetic field located near the environments border. The resonant mode of the mutual influence of COTS and the resonance of the dye molecules lead to the split of the COTS frequency. The splitting effect depends significantly on the concentration of dye molecules as well as on the step of the CLC helix. When reconstructing from the resonance, the splitting effect of COTS is persistent as long as the COTS lies in the frequency dispersion of the dye molecules.

2. Description of the Model

The investigated finite structure schematically shown in Figure 1a consists of a PPAM conjugated with a right-handed DDCLC. The multilayer PPAM structure consists, in its turn, of alternating uniaxial dielectric layers with different refractive indices $n_e^p = \sqrt{\varepsilon_e^p}$ and $n_o^p = \sqrt{\varepsilon_o^p}$ and it is characterized by dielectric tensors of two neighboring layers, which can be written as:

$$\hat{\varepsilon}_V = \begin{pmatrix} 1.7 & 0 & 0 \\ 0 & 1.5 & 0 \\ 0 & 0 & 1.7 \end{pmatrix}, \quad \hat{\varepsilon}_H = \begin{pmatrix} 1.5 & 0 & 0 \\ 0 & 1.7 & 0 \\ 0 & 0 & 1.7 \end{pmatrix}. \tag{1}$$

In should be noted what the optical axes of the PPAM are directed along x (yellow layer) and y (grey layer) axis respectively (see Figure 1a). The number of unit cells (the number of V-H pairs) in the structure is N_{PPAM}, the period of the structure is $\Lambda = d_V + d_H$, where d_V and d_H are the thicknesses of the unit cell layers; and the PPAM thickness is $d = N_{PPAM} \times \Lambda$. The CLCs with a continuous helical symmetry of the permittivity tensor are characterized by the helix pitch p, the cholesteric layer thickness L, the number of periods N_{CLC}, and the ordinary and extraordinary refractive indices $n_\parallel = n_e = \sqrt{\varepsilon_\parallel}$ and $n_\perp = n_o = \sqrt{\varepsilon_\perp}$. The structure under study is limited to a medium with the refractive index $(n_e + n_o)/2$.

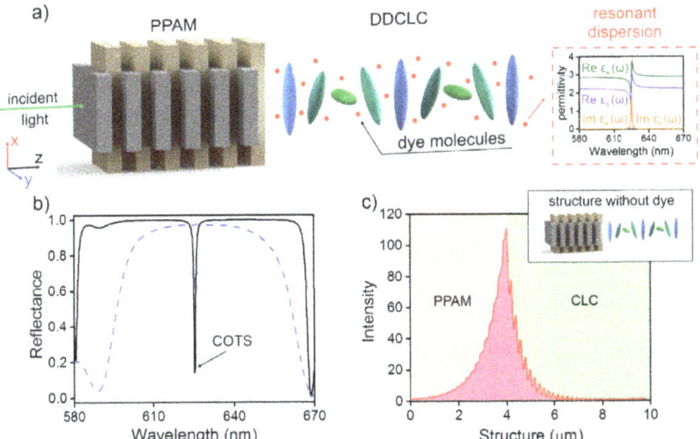

Figure 1. (**a**) Schematic of the structure consisting of a PPAM with length d and a DDCLC with length L. Inset: imaginary (yellow curve) and real (green and purple curves) are parts of the effective permittivity component of the DDCLC tensor. (**b**) Reflectance spectra of the PPAM-CLC structure. The solid curve corresponds to the diffracting polarization, i.e., the circular polarization that is reflected from CLC and the dashed curve, to the nondiffracting one. The PPAM length is $d = 3.92$ μm, the period is $\Lambda = d_V + d_H$, $d_V = 100$ nm, $d_H = 96$ nm, the number of periods is $N_{PPAM} = 20$, and the refractive indices are $n_e^p = 1.7$ and $n_o^p = 1.5$. The CLC layer length is $L = 5.85$ μm, the number of periods is $N_{CLC} = 30$, the helix pitch is $p = 390$ nm, and the extraordinary and ordinary refractive indices are $n_e = 1.7$ and $n_o = 1.5$. The CLC band gap center is $\lambda_0 = 625$ nm and $\phi = \pi/4$. (**c**) Spatial distribution of the local field intensity in the sample corresponding to a COTS wavelength of $\lambda = 625.3$ nm normalized to the initial value.

We consider the normal incidence of light onto the structure. The angle between the CLC director and the optical axis of the PPAM layer conjugated with a CLC is denoted by φ. If light propagates along the axis, the CLC permittivity and permeability tensors are:

$$\hat{\varepsilon}(z) = \varepsilon_m \begin{pmatrix} 1 + \delta\cos(qz) & \pm\delta\sin(2qz) & 0 \\ \pm\delta\sin(qz) & 1 - \delta\cos(qz) & 0 \\ 0 & 0 & 1 - \delta \end{pmatrix}, \quad \hat{\mu}(z) = \hat{I}, \tag{2}$$

respectively, where $q = 4\pi/p$, $\varepsilon_m = (\varepsilon_e + \varepsilon_o)/2$ and $\delta = (\varepsilon_e - \varepsilon_o)/(\varepsilon_e + \varepsilon_o)$.

It is supposed that the order parameter that characterizes the degree of the dipole moment order of the dye molecules transition is zero. This corresponds to the chaotic orientation of the dipole moment of the dye molecules transition. In this case, the presence of dye molecules in the CLC matrix causes the frequency dependence of the main values of the local dielectric tensor, and we assume the Lorentz form of this frequency dependence [37]. In addition, a small volume concentration of the dye molecules in the CLC matrix is suggested, since the coupling of oscillators through a local field is not taken into account. The characteristic concentrations of the dye molecules correspond to the concentration of molecules in an ideal gas under normal conditions. Lorentz form of their frequency dependence:

$$\varepsilon_e(\omega) = \varepsilon_e + \frac{f_1}{\omega_{01}^2 - \omega^2 - i\gamma_1\omega},$$ (3)

$$\varepsilon_o(\omega) = \varepsilon_o + \frac{f_2}{\omega_{02}^2 - \omega^2 - i\gamma_2\omega},$$ (4)

where γ_1 and γ_2 are the damping coefficients, ω_{01} and ω_{02} are the resonant frequencies, f_1 and f_2 are the quantities proportional to the oscillator strength $\overline{f}_{1,2}$.

$$f_{1,2} = \frac{4\pi Ne^2}{m} \times \overline{f}_{1,2}.$$ (5)

In Equation (5), N is the number of dye molecules in the unit volume; the dipole oscillator strength $\overline{f}_{1,2}$ is, as a rule, about tenths of unity; and e and m are the electron charge and mass, respectively. Hereinafter in the calculation, we assume $f_1 = f_2 = f$, $\gamma_1 = \gamma_2 = \gamma$, $\omega_{01} = \omega_{02} = \omega_0$.

The possible way of fabrication of the proposed structure is to use polymer-stabilized liquid crystals [35,39–41]. Dye-doped polymeric cholesteric liquid crystal films can be made in the same manner as Shmidtke et al. [35] and Jeong et al. [41] who studied defect mode lasing in such films. PPAM can be created from polymerized nematic layers, for example 5CB. Each layer of LC should be applied sequentially and polymerized. Despite the use of polymerized materials, the possibility of manipulating CLCs remains [42,43].

A numerical analysis of the spectral properties of the system and the field distribution in the sample of a dye-doped cholesteric conjugated with an anisotropic mirror is performed using the Berreman 4 × 4 transfer matrix method [44]. The equation describing the propagation of light at frequency ω along the z axis normal to the structural layers has the form:

$$\frac{d\psi}{dz} = \frac{i\omega}{c}\Delta(z)\psi(z),$$ (6)

where $\psi(z) = (E_x, H_y, E_y, -H_x)^T$ and $\Delta(z)$ is the Berreman matrix, which depends on the dielectric function and the incident wave vector.

3. Results and Discussion

Figure 1b,c, show the reflectance spectra and the local field intensity distribution at wavelength of COTS in the PPAM-CLC sample without dye molecules. It should be noted that, at the equal PPAM and CLC refractive indices, the CLC band gap lies in the wavelength range of 580–670 nm, which is 20 nm higher than the PPAM band gap with the boundaries from 595 to 665 nm. It is well-known that the complete or partial overlap of the band gaps in the energy spectrum is a condition for the formation of a COTS at the interface between two media. The fulfillment of this condition ensures the excitation of a localized state at the diffracting polarization at a wavelength of $\lambda = 625.3$ nm, while at the polarization of the opposite sign (nondiffracting polarization), the COTS is not excited. The observed state is high-Q and can be effectively tuned in frequency [30].

We add the CLC with fluorescent dye molecules, which have parameters of $\gamma = 4 \times 10^{12}$ s^{-1} and $f = 2 \times 10^{28}$ s^{-2}. In this case, the principal values of the CLC local dielectric tensor become frequency-dependent. If the resonant frequency of dye molecules coincides with the COTS frequency, then two modes, instead of one, appear in the spectra. These modes yield dips at wavelengths of $\lambda_1 = 619.3$ nm and $\lambda_2 = 631.3$ nm (see Figure 2). The splitting value is $\Delta\lambda = 12$ nm. At an oscillator strength of $\overline{f} \approx 0.5$, Equation (5) yields an estimated number of $N \approx 10^{19}$ cm^{-3} of dye molecules in the unit volume.

Figure 2d shows the spatial distributions of the COTS local field intensity at wavelengths of 619.3 and 631.3 nm, which correspond to the reflection maxima in the spectrum (Figure 2a). It can be seen that the light is localized at the PPAM-DDCLC interface and the local field intensity decreases exponentially with increasing distance from the interface. There are two different kinds of the local intensity rippling inside PPAM and DDCLC.

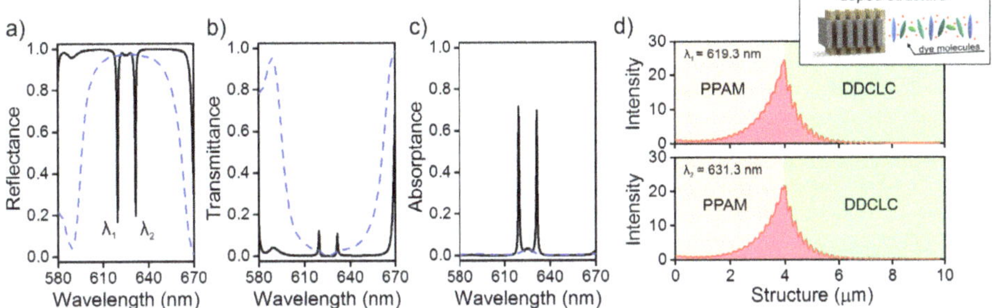

Figure 2. (**a**) Reflectance, (**b**) transmittance, and (**c**) absorptance spectra of the PPAM-DDCLC structure. The solid curve corresponds to the diffracting polarization and the dashed curve, to the nondiffracting one. The damping parameters are $\gamma = 4 \times 10^{12}$ s^{-1} and $f = 2 \times 10^{28}$ s^{-2}. The rest parameters are the same as in Figure 1b. (**d**) Spatial distribution of the local field intensity in the sample normalized to the initial value.

First, the PPAM rippling is due to nonhomogeneity of PPAM material. The layers are virtually quarter-vawelength. The multiple layer boundaries correspond to standing wave nodes and antinodes of x and y components of electric field [30], that results in rippling of the overall field. Second, the local intensity rippling in CLC has different origin, because the CLC material is homogeneously twisted along z direction and its main eigenwave is a smoothly evanescenting exponential [20]. In total there are four eigenwaves inside the CLC, and all of them are excited with the amplitudes depending on boundary conditions and excitation wave polarization. The interference of eigenwaves results in some rippling, small minima and maxima of the overall field.

The Q factor of the two modes obtained by splitting, at the resonant frequency of dye molecules coinciding with the COTS center, will be approximately the same and equal to 774 for the high-frequency mode and 702 for the low-frequency mode. At the detuning of the resonant frequency from the COTS wavelength, one can enhance the Q factor of one peak by reducing the Q factor of the other. Below, we show that the Q factor of the new COTSs obtained by the splitting can be significantly increased by changing the parameters of dye molecules.

Let us consider the effect of the dye molecule concentration on the splitting value and the COTS position. Figure 3a–c show that, with an increase in the dye molecule concentration by an order of magnitude, two COTSs arise at wavelengths of 607 and 644.9 nm, respectively, and the splitting value increases by 25.9 nm and attains $\Delta\lambda = 37.9$ nm. As the concentration of fluorescent molecules decreases by an order of magnitude, the COTSs are observed at wavelengths of 623.4 and 627.1 nm and the splitting value decreases down to $\Delta\lambda = 3.7$ nm.

Figure 3d–f illustrate the sensitivity of the COTS splitting to the changes in the damping coefficient. We do not specify the physical origins of the damping, which can be very different. As the γ value decreases by two times, the COTSs appear in the spectra at wavelengths of 619.3 and 631.3 nm and the Q factor of the split modes grows to about 1000. In this case, the splitting value remains the same and amounts to $\Delta\lambda = 12$ nm. As the γ value increases by an order of magnitude, the new COTSs become poorly distinguishable, the peak maxima corresponding to wavelengths of 620.2 and 630.3 nm approach each other, and the splitting value decreases to $\Delta\lambda = 10.1$ nm. We would like to note that, at $\gamma = 4 \times 10^{14}$ s^{-1}, the splitting effect does not manifest itself, and the low-Q dip at a wavelength of 625.3 nm is observed in the reflectance spectrum, which corresponds to the COTS wavelength in the PPAM-CLC structure without dye.

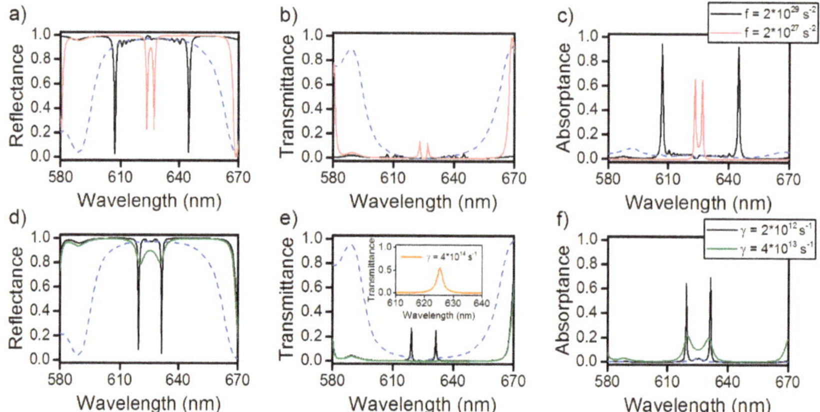

Figure 3. (**a,d**) Reflectance, (**b,e**) transmittance, and (**c,f**) absorptance spectra of the investigated structure at different volume concentrations of dye molecules and damping coefficients. Solid curves correspond to the diffracting polarization and the dashed curve, to the nondiffracting one. The rest parameters are the same as in Figure 2. The insert in figure (**e**) shows the transmission spectra of the structure for $\gamma = 4 \times 10^{14}\ \mathrm{s}^{-1}$.

As was mentioned above, an important advantage of the CLCs over other types of photonic crystals is their high sensitivity to external fields. A strong dependence of the helix pitch, for example, on temperature or applied voltage can be used to effectively control the COTS splitting value. Thus, a decrease in the helix pitch by 10 nm leads to an increase in the splitting value by 1.2 nm to $\Delta\lambda = 13.2$ nm (Figure 4a–c). In this case, as can be seen in Figure 4a,c, the positions of the COTS frequencies shift to the short-wavelength spectral region by 615.6 and 628.8 nm, respectively. In the transmittance spectrum shown in Figure 4b, due to the damping, the only short-wavelength localized peak at a wavelength of 615.6 nm remains. With an increase in the helix pitch by 10 nm, the opposite situation is implemented. It can be seen from the reflectance and absorptance spectra shown in Figure 4a,c that, with an increase in the helix pitch to 400 nm, the positions of the COTS frequencies shift to the long-wavelength spectral region to 621.7 and 635.5 nm and the splitting value increases to $\Delta\lambda = 13.5$ nm. In the transmittance spectrum in Figure 4b, the only long-wavelength COTS remains localized at a wavelength of 635.5 nm.

Figure 4d–f illustrate the possibility of controlling the spectral properties of the COTS by optimizing the geometric parameters of the PPAM-DDCLC structure. Let us reduce the size of the structure by decreasing the N_{PPAM} value from 20 to 15 and the N_{CLC} value from 30 to 20. In this case, the spectral positions of the COTS frequencies do not change, i.e., the splitting value remains unchanged, while the Q factor of the peaks decreases. At the increasing sample length, at $N_{CLC} = 40$ and $N_{PPAM} = 25$, due to the increased band gap contrast, the gaps in the reflectance spectrum are about 38% and the corresponding COTS peaks in the transmittance spectrum are almost absent, due to an increase in the length of examined sample.

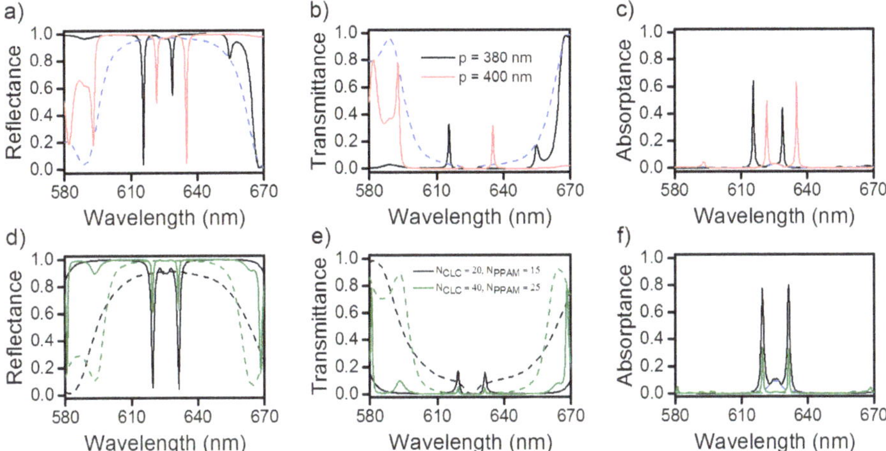

Figure 4. (**a,d**) Reflectance, (**b,e**) transmittance, and (**c,f**) absorptance spectra of the investigated structure at different cholesteric helix pitches and PPAM and CLC geometric parameters. Solid lines correspond to the diffracting polarization and the dashed lines, to the nondiffracting one. The rest parameters are the same as in Figure 2.

The similar effects can be implemented in a different way, i.e., by changing the angle φ between the PPAM and DDCLC optical axes at the interface between the media (inset in Figure 5). When passing the interface between two mirrors, the geometric phase is controlled by rotation of the mirrors in the interface plane. Figure 5a,b show that, in the reflectance and transmittance spectra, at the diffracting polarization, depending on the angle φ, the reflection of the right-hand peak decreases from 24% to 8%, while the transmittance increases from 11% to 50%, with an increase in ϕ from $\pi/4$ to $\pi/3$. Upon further rotation of the angle, the transmittance of the long-wave peak decreases and the reflectance increases. We would like to note that, when ϕ is rotated in the opposite direction, the inverse situation is observed in the transmittance spectrum, when the transmittance is only preserved for the short-wave COTS.

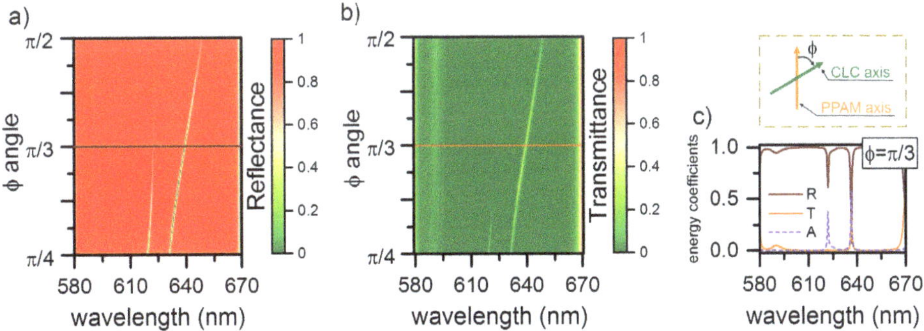

Figure 5. (**a**) Reflectance and (**b**) transmittance spectra of the structure at different angles φ. (**c**) Reflectance, transmittance, and absorptance spectra of the structure at $\varphi = \pi/3$. Insert: Definition of angle ϕ as the angle between the CLC director at the boundary and the optical axis of the PPAM layer conjugated with the CLC. The rest parameters are the same as in Figure 2.

Figure 5c shows the reflectance, transmittance, and absorptance spectra of the structure at $\varphi = \pi/3$. In this case, the COTSs appear in the spectra at wavelengths of 622 and 636.4 nm.

The splitting value increases by 4.4 nm to $\Delta\lambda = 16.4$ nm and the transmittance spectrum only reflects the long-wavelength COTS. Note that when changing the helix pitch or angle φ, the position of the COTS slightly deviates from the value of 625.3 nm. In this case, the splitting effect will be preserved only if the frequency of COTS lies in the frequency dispersion range of the dye molecules.

4. Conclusions

Thus, the study of the properties of the PPAM-CLC-based model structures shows that doping of a CLC with dye molecules leads to the splitting of the COTS localized at the PPAM-DDCLC interface, if the resonant frequency of dye molecules coincides with the COTS frequency. As a result, the resonances corresponding to the two modes localized at the interface appear in the spectra. The possibility of effective controlling the spectral position and Q factor of these resonances, as well as the COTS splitting value, by changing the dye parameters and the CLC and PPAM geometric parameters was demonstrated. The proposed structure can be used to create miniature lasers with the circularly polarized fundamental mode, as well as to design narrow-band and tunable filters.

Author Contributions: A.Y.A.—performed the calculations and drafted the manuscript; R.G.B.—visualized the results; M.V.P. and N.V.R.—helped with the software and methods. I.V.T. and S.Y.V. supervisors the whole study and finalized the manuscript. All authors have read and agreed to the submitted version of the manuscript.

Funding: The reported study was funded by Russian Foundation for Basic Research, Government of Krasnoyarsk Territory, Krasnoyarsk Region Science and Technology Support Fund to the research project No. 19-42-240004 and by Russian Foundation for Basic Research, project No. 19-52-52006

Acknowledgments: The authors are thankful to Pavel S. Pankin for valuable discussions and comments.

Conflicts of Interest: The authors declare no conflict of interest.

Abbreviations

The following abbreviations are used in this manuscript:

OTS	Optical Tamm State
COTS	Chiral Optical Tamm State
PPAM	Polarization-Preserving Anisotropic Mirror
CLC	Cholesteric Liquid Crystal
DDCLC	Dye-doped Cholesteric Liquid Crystal

References

1. Kavokin, A.V.; Shelykh, I.A.; Malpuech, G. Lossless interface modes at the boundary between two periodic dielectric structures. *Phys. Rev. B* **2005**, *72*, doi:10.1103/physrevb.72.233102. [CrossRef]
2. Kaliteevski, M.; Iorsh, I.; Brand, S.; Abram, R.A.; Chamberlain, J.M.; Kavokin, A.V.; Shelykh, I.A. Tamm plasmon-polaritons: Possible electromagnetic states at the interface of a metal and a dielectric Bragg mirror. *Phys. Rev. B* **2007**, *76*, doi:10.1103/physrevb.76.165415. [CrossRef]
3. Vinogradov, A.P.; Dorofeenko, A.V.; Merzlikin, A.M.; Lisyansky, A.A. Surface states in photonic crystals. *Phys. Uspekhi* **2010**, *53*, 243–256. [CrossRef]
4. Sasin, M.E.; Seisyan, R.P.; Kalitteevski, M.A.; Brand, S.; Abram, R.A.; Chamberlain, J.M.; Egorov, A.Y.; Vasil'ev, A.P.; Mikhrin, V.S.; Kavokin, A.V. Tamm plasmon polaritons: Slow and spatially compact light. *Appl. Phys. Lett.* **2008**, *92*, 251112. [CrossRef]
5. Goto, T.; Dorofeenko, A.V.; Merzlikin, A.M.; Baryshev, A.V.; Vinogradov, A.P.; Inoue, M.; Lisyansky, A.A.; Granovsky, A.B. Optical Tamm States in One-Dimensional Magnetophotonic Structures. *Phys. Rev. Lett.* **2008**, *101*. [CrossRef] [PubMed]
6. Symonds, C.; Lheureux, G.; Hugonin, J.P.; Greffet, J.J.; Laverdant, J.; Brucoli, G.; Lemaitre, A.; Senellart, P.; Bellessa, J. Confined Tamm Plasmon Lasers. *Nano Lett.* **2013**, *13*, 3179–3184. [CrossRef]
7. Jiménez-Solano, A.; Galisteo-López, J.F.; Míguez, H. Light-Emitting Coatings: Flexible and Adaptable Light-Emitting Coatings for Arbitrary Metal Surfaces based on Optical Tamm Mode Coupling (Advanced Optical Materials 1/2018). *Adv. Opt. Mater.* **2018**, *6*, 1870001. [CrossRef]

8. Yang, Z.Y.; Ishii, S.; Yokoyama, T.; Dao, T.D.; Sun, M.G.; Pankin, P.S.; Timofeev, I.V.; Nagao, T.; Chen, K.P. Narrowband Wavelength Selective Thermal Emitters by Confined Tamm Plasmon Polaritons. *ACS Photonics* **2017**, *4*, 2212–2219. [CrossRef]

9. Wang, X.; Liang, Y.; Wu, L.; Guo, J.; Dai, X.; Xiang, Y. Multi-channel perfect absorber based on a one-dimensional topological photonic crystal heterostructure with graphene. *Opt. Lett.* **2018**, *43*, 4256. [CrossRef]

10. Bikbaev, R.; Vetrov, S.; Timofeev, I. Epsilon-Near-Zero Absorber by Tamm Plasmon Polariton. *Photonics* **2019**, *6*, 28. [CrossRef]

11. Huang, S.G.; Chen, K.P.; Jeng, S.C. Phase sensitive sensor on Tamm plasmon devices. *Opt. Mater. Express* **2017**, *7*, 1267. [CrossRef]

12. Bikbaev, R.G.; Vetrov, S.Y.; Timofeev, I.V. Hybrid Tamm and surface plasmon polaritons in resonant photonic structure. *J. Quant. Spectrosc. Radiat. Transf.* **2020**, *253*, 107156. [CrossRef]

13. Zhang, X.L.; Song, J.F.; Li, X.B.; Feng, J.; Sun, H.B. Optical Tamm states enhanced broad-band absorption of organic solar cells. *Appl. Phys. Lett.* **2012**, *101*, 243901. [CrossRef]

14. Lu, H.; Li, Y.; Yue, Z.; Mao, D.; Zhao, J. Topological insulator based Tamm plasmon polaritons. *APL Photonics* **2019**, *4*, 040801. [CrossRef]

15. Wang, L.; Cai, W.; Bie, M.; Zhang, X.; Xu, J. Zak phase and topological plasmonic Tamm states in one-dimensional plasmonic crystals. *Opt. Express* **2018**, *26*, 28963. [CrossRef]

16. Jeng, S.C. Applications of Tamm plasmon-liquid crystal devices. *Liq. Cryst.* **2020**, 1–9. [CrossRef]

17. Cheng, H.C.; Kuo, C.Y.; Hung, Y.J.; Chen, K.P.; Jeng, S.C. Liquid-Crystal Active Tamm-Plasmon Devices. *Phys. Rev. Appl.* **2018**, *9*. [CrossRef]

18. Ferrier, L.; Nguyen, H.S.; Jamois, C.; Berguiga, L.; Symonds, C.; Bellessa, J.; Benyattou, T. Tamm plasmon photonic crystals: From bandgap engineering to defect cavity. *APL Photonics* **2019**, *4*, 106101. [CrossRef]

19. Adams, M.; Cemlyn, B.; Henning, I.; Parker, M.; Harbord, E.; Oulton, R. Model for confined Tamm plasmon devices. *J. Opt. Soc. Am. B* **2018**, *36*, 125. [CrossRef]

20. Belyakov, V. *Diffraction Optics of Complex-Structured Periodic Media*; Springer International Publishing: Cham, Switzerland, 2019. [CrossRef]

21. Dolganov, P.V.; Ksyonz, G.S.; Dolganov, V.K. Photonic liquid crystals: Optical properties and their dependence on light polarization and temperature. *Phys. Solid State* **2013**, *55*, 1101–1104. [CrossRef]

22. Vetrov, S.Y.; Pyatnov, M.V.; Timofeev, I.V. Surface modes in "photonic cholesteric liquid crystal–phase plate–metal" structure. *Opt. Lett.* **2014**, *39*, 2743. [CrossRef] [PubMed]

23. Timofeev, I.V.; Vetrov, S.Y. Chiral optical Tamm states at the boundary of the medium with helical symmetry of the dielectric tensor. *JETP Lett.* **2016**, *104*, 380–383. [CrossRef]

24. Rudakova, N.V.; Timofeev, I.V.; Vetrov, S.Y.; Lee, W. All-dielectric polarization-preserving anisotropic mirror. *OSA Contin.* **2018**, *1*, 682. [CrossRef]

25. Plum, E.; Zheludev, N.I. Chiral mirrors. *Appl. Phys. Lett.* **2015**, *106*, 221901. [CrossRef]

26. Reusch, E. Untersuchung über Glimmercombinationen. *Ann. Der Phys. Und Chem.* **1869**, *214*, 628–638. [CrossRef]

27. Joly, G.; Isaert, N. Some electromagnetic waves in Reusch's piles. IV. Multiple domains of selective reflection. *J. Opt.* **1986**, *17*, 211–221. [CrossRef]

28. Dixit, M.; Lakhtakia, A. Selection strategy for circular-polarization-sensitive rejection characteristics of electro-optic ambichiral Reusch piles. *Opt. Commun.* **2008**, *281*, 4812–4823. [CrossRef]

29. Hodgkinson, I.J.; Lakhtakia, A.; Wu, Q.; Silva, L.D.; McCall, M.W. Ambichiral, equichiral and finely chiral layered structures. *Opt. Commun.* **2004**, *239*, 353–358. [CrossRef]

30. Rudakova, N.V.; Timofeev, I.V.; Bikbaev, R.G.; Pyatnov, M.V.; Vetrov, S.Y.; Lee, W. Chiral Optical Tamm States at the Interface between an All-Dielectric Polarization-Preserving Anisotropic Mirror and a Cholesteric Liquid Crystal. *Crystals* **2019**, *9*, 502. [CrossRef]

31. Pyatnov, M.V.; Timofeev, I.V.; Vetrov, S.Y.; Rudakova, N.V. Coupled chiral optical Tamm states in cholesteric liquid crystals. *Photonics* **2018**, *5*, 30. [CrossRef]

32. Hernández, J.C.; Reyes, J.A. Optical band gap in a cholesteric elastomer doped by metallic nanospheres. *Phys. Rev. E* **2017**, *96*. [CrossRef]

33. Lin, J.D.; Hsieh, M.H.; Wei, G.J.; Mo, T.S.; Huang, S.Y.; Lee, C.R. Optically tunable/switchable omnidirectionally spherical microlaser based on a dye-doped cholesteric liquid crystal microdroplet with an azo-chiral dopant. *Opt. Express* **2013**, *21*, 15765. [CrossRef] [PubMed]

34. Ilchishin, I.; Tikhonov, E.; Tishchenko, V.; Shpak, M. Generation of tunable radiation by impurity cholesteric liquid crystals. *J. Exp. Theor. Phys. Lett.* **1980**, *32*, 24–27.

35. Schmidtke, J.; Stille, W.; Finkelmann, H. Defect Mode Emission of a Dye Doped Cholesteric Polymer Network. *Phys. Rev. Lett.* **2003**, *90*. [CrossRef] [PubMed]

36. Risse, A.M.; Schmidtke, J. Angular-dependent spontaneous emission in cholesteric liquid-crystal films. *J. Phys. Chem. C* **2019**, *123*, 2428–2440. [CrossRef]

37. Gevorgyan, A.H. Fano resonance in a cholesteric liquid crystal with dye. *Phys. Rev. E* **2019**, *99*. [CrossRef]

38. Vetrov, S.Y.; Timofeev, I.V.; Shabanov, V.F. Localized modes in chiral photonic structures. *Phys. Uspekhi* **2020**, *63*, 33–56. [CrossRef]

39. Song, M.H.; Park, B.; Shin, K.C.; Ohta, T.; Tsunoda, Y.; Hoshi, H.; Takanishi, Y.; Ishikawa, K.; Watanabe, J.; Nishimura, S.; et al. Effect of phase retardation on defect-mode lasing in polymeric cholesteric liquid crystals. *Adv. Mater.* **2004**, *16*, 779–783. [CrossRef]

40. Mitov, M.; Dessaud, N. Going beyond the reflectance limit of cholesteric liquid crystals. *Nat. Mater.* **2006**, *5*, 361–364. [CrossRef]

41. Jeong, S.M.; Ha, N.Y.; Takanishi, Y.; Ishikawa, K.; Takezoe, H.; Nishimura, S.; Suzaki, G. Defect mode lasing from a double-layered dye-doped polymeric cholesteric liquid crystal films with a thin rubbed defect layer. *Appl. Phys. Lett.* **2007**, *90*, 261108. [CrossRef]

42. Menzel, A.M.; Brand, H.R. Cholesteric elastomers in external mechanical and electric fields. *Phys. Rev. E* **2007**, *75*, 011707. [CrossRef] [PubMed]

43. Nagai, H.; Urayama, K. Thermal response of cholesteric liquid crystal elastomers. *Phys. Rev. E* **2015**, *92*, 022501. [CrossRef] [PubMed]

44. Berreman, D.W. Optics in Stratified and Anisotropic Media: 4x4-Matrix Formulation. *J. Opt. Soc. Am.* **1972**, *62*, 502. [CrossRef]

Article

Chiral-Selective Tamm Plasmon Polaritons

Meng-Ying Lin [1,2], Wen-Hui Xu [1,2], Rashid G. Bikbaev [3,4,*], Jhen-Hong Yang [5,6], Chang-Ruei Li [7,8], Ivan V. Timofeev [3,4], Wei Lee [1,2] and Kuo-Ping Chen [1,2,*]

1   Institute of Imaging and Biomedical Photonics, College of Photonics, National Chiao Tung University, 301 Gaofa 3rd Road, Tainan 71150, Taiwan; e1253400@gmail.com (M.-Y.L.); uxxsh8@gmail.com (W.-H.X.); wlee@nctu.edu.tw (W.L.)
2   Institute of Imaging and Biomedical Photonics, College of Photonics, National Yang Ming Chiao Tung University, 301 Gaofa 3rd Road, Tainan 71150, Taiwan
3   Kirensky Institute of Physics, Federal Research Center KSC SB RAS, 660036 Krasnoyarsk, Russia; ivan-v-timofeev@ya.ru
4   Siberian Federal University, 660041 Krasnoyarsk, Russia
5   Institute of Photonic System, College of Photonics, National Chiao Tung University, 301 Gaofa 3rd Road, Tainan 71150, Taiwan; s87069@hotmail.com
6   Institute of Photonic System, College of Photonics, National Yang Ming Chiao Tung University, 301 Gaofa 3rd Road, Tainan 71150, Taiwan
7   Institute of Lighting and Energy Photonics, College of Photonics, National Chiao Tung University, 301 Gaofa 3rd Road, Tainan 71150, Taiwan; g10592000@hotmail.com
8   Institute of Lighting and Energy Photonics, College of Photonics, National Yang Ming Chiao Tung University, 301 Gaofa 3rd Road, Tainan 71150, Taiwan
*   Correspondence: bikbaev@iph.krasn.ru (R.G.B.); kpchen@nctu.edu.tw (K.-P.C.)

Abstract: Chiral-selective Tamm plasmon polariton (TPP) has been investigated at the interface between a cholesteric liquid crystal and a metasurface. Different from conventional TPP that occurs with distributed Bragg reflectors and metals, the chiral–achiral TPP is successfully demonstrated. The design of the metasurface as a reflective half-wave plate provides phase and polarization matching. Accordingly, a strong localized electric field and sharp resonance are observed and proven to be widely tunable.

Keywords: metasurfaces; tamm plasmon polaritons; chirality

Citation: Lin, M.-Y.; Xu, W.-H.; Bikbaev, R.G.; Yang, J.-H.; Li, C.-R.; Timofeev, I.V.; Lee, W.; Chen, K.-P. Chiral-Selective Tamm Plasmon Polaritons. *Materials* **2021**, *14*, 2788. https://doi.org/10.3390/ma14112788

Academic Editor: Maria Principe

Received: 23 April 2021
Accepted: 21 May 2021
Published: 24 May 2021

Publisher's Note: MDPI stays neutral with regard to jurisdictional claims in published maps and institutional affiliations.

1. Introduction

In recent years, metamaterials have been widely utilized in photoelectronics due to the advances in controlling the phase, polarizations, and chirality. Indeed, chirality gives an additional degree of freedom in photonic systems. Therefore, chiral photonics has received a lot of attention lately, such as chiral-selective metamirrors [1–4], chiral quantum optics [5], spectropolarimetry [6], etc. Chiral properties can be effectively enhanced using metamaterials and photonic-crystal cavities. In the literature, chirality surface states could be observed on the surface of topological materials [7] or at the interface of two cholesteric liquid crystals (CLCs) [8,9]. However, it would be difficult to be observed at a chiral–achiral interface as the polarization state could not be preserved. An example of such an interface state is Tamm plasmon polariton (TPP). It was first proposed in 2007 [10] and is similar to the Tamm state in a semiconductor, where electrons are localized at the surface of the crystal. The TPP appears between the metal and the periodic dielectric of high and low refractive indices, which is called a distributed Bragg reflector (DBR). Later it was shown that this state can be utilized for absorbers [11], sensors [12–14], Tamm plasmon lasers [15–17], and solar cells [18,19]. The excitation of chiral-selective TPP at the interface between a CLC and a flat metal film is impossible (please see Figure 1a,b), and the resonance dip cannot be seen within the CLC stopband [20] unless the polarization of reflected light from the metal is changed. Then, the high reflection of the CLC stopband can be maintained without the localization of the light at the interface between the CLC and the metal.

In this regard, a novel design combining a CLC and a half-wave plate metasurface is proposed. The possibility of excitation of a chiral TPP in this structure is demonstrated experimentally for the first time and confirmed numerically. The tuning of the chiral-TPP wavelength is shown by varying the temperature.

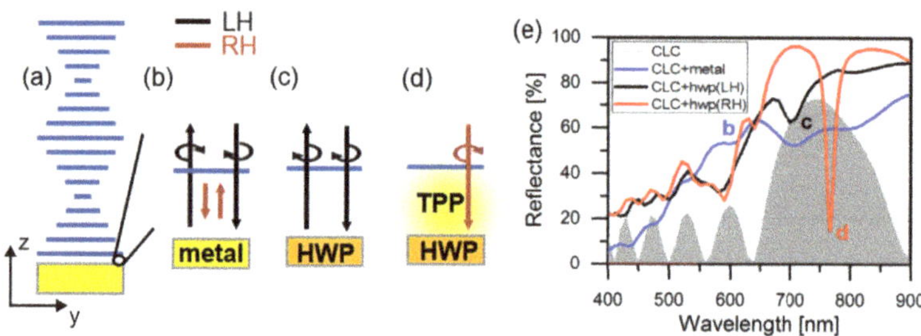

Figure 1. (**a**) Schematic of the right-handed helical CLC combined with a bottom film. Polarization dynamics of the light between (**b**) CLC and a metal mirror and (**c**,**d**) CLC and a half-wave plate metasurfaces. (**e**) Reflectance spectra of CLC–metal and CLC-half-wave plate simulated by the software for Multiphysics simulation COMSOL. The rotation arrow direction indicates the right-handed (RH) and the left-handed (LH) circular polarizations propagating along the z axis. The center of the CLC stopband is at 750 nm. The refractive index of the SiO_2 layer is fixed at 1.45, and that of the gold nanobrick and PMMA layer comes from the databases of Johnson and Christy [21] and Sultanova [22], respectively. The full reflectance spectra of the measurements is presented in the Supplementary Materials.

2. Description of the Model

Figure 1b shows that partial right-handed circularly polarized light is transmitted through the right-handed CLC layer of a finite thickness, and the polarization changes when reflected from the metal. The reflected light with left-handed circular polarization passes through the CLC, and the reflectance spectra correspond to the blue line in Figure 1e, resembling the combination of reflection spectra of only metal and CLC. In order to preserve the reflected circular polarization, a quarter-wave plate was proposed to match the phase [23]. As the partial right-handed circular polarization transforms into linear polarization when light passes through the quarter-wave plate, the polarization of the reflected light remained unchanged. Hence, right-handed circularly polarized light is localized between the CLC and metal, yielding resonance. However, in reality, a typical quarter-wave plate is much thicker than the wavelengths, which could not sustain the surface waves. For example, a 75 μm thick conventional quarter-wave plate (Edmund Optics) has been tested and it was found that the interference diminishes the TPP resonance. Here, we replace the thick phase plate by a metasurface, with the function of a reflective half-wave plate (HWP), as shown in Figure 2. The dimensions of the unit cell with a nanobrick fabricated using electron beam lithography are shown (please see Supplementary Materials). The SiO_2 layer and bottom metal film are used to control the phase of the reflected light [24–29]. This design allows the handedness of reflected light to be preserved and the polarization-matching condition fulfilled. In this paper, the experimentally measured Q factor of the chiral-selective TPP is 27.2. The Q factor of the TPP resonance is conventionally determined by the losses and the volume of the resonator. In our case, the cavity volume is mainly governed by the CLC thickness. In addition, loss is predominantly due to the absorption in the plasmonic metasurface. Alternatively, the all-dielectric structures allowed us to obtain higher Q values at the expense of compactness. Additionally, the experimental implementation of all-dielectric handedness-preserving structures is difficult [30].

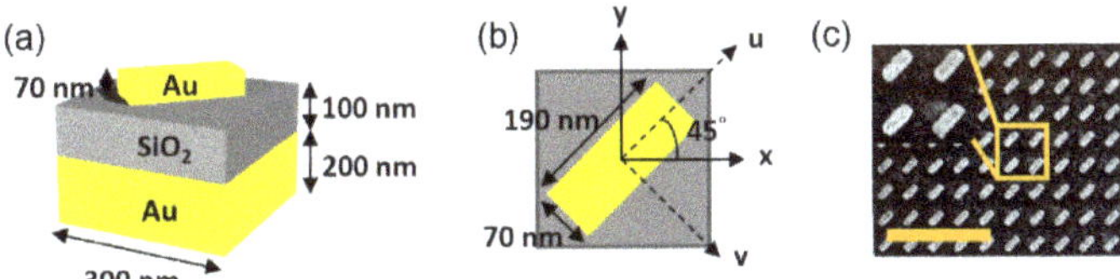

Figure 2. (**a**,**b**) Schematic of a unit cell of metasurface consisting of an Au-SiO$_2$-Au structure. The dashed axes of "u" and "v" define the longer and shorter axes of the nanobrick. The "x" axis defines the CLC director at the surface; it is oriented at $\chi = 45°$. (**c**) The scanning electron microscopic image (the scale bar is 1.2 μm). The detailed images of the nanobricks are shown in the inset.

The phase matching is another crucial condition for localized state excitation. This second condition is equivalent to the geometric phase condition for the angle between the long axis of the metasurface nanobrick and the CLC direction [31,32]. In Figure 2b, this angle is shown to be $\chi = 45°$ and the TPP resonance frequency approximately corresponds to the center of the CLC stopband. By varying, the frequency can be easily tuned through the entire CLC stopband and even switched off [31]. This possibility is provided by the junction of two mirrors with unique and complementary properties. The CLC is chiral and the metasurface is achiral hence anisotropic. The CLC is fluidic and tunable and the metasurface serves as a robust solid basement for tunability. Such chiral tunability is intensively investigated in self-organized structures related to tensagrity, durotaxis and phototropism [33,34].

3. Results

The phase difference in reflection is defined as the phase of the u-polarized light subtracting the phase of the v-polarized light (Figure 3a). It is assumed that the light is incident on the metasurface from the air. The phase difference approaches π in the wavelength range from 700 to 900 nm, which satisfies the properties of a half-wave plate [35–37]. Due to the phase change, the direction of the rotation is opposite when circularly polarized light impinges on the half-wave plate. Therefore, the handedness remains unchanged between the reflected light and the incident light.

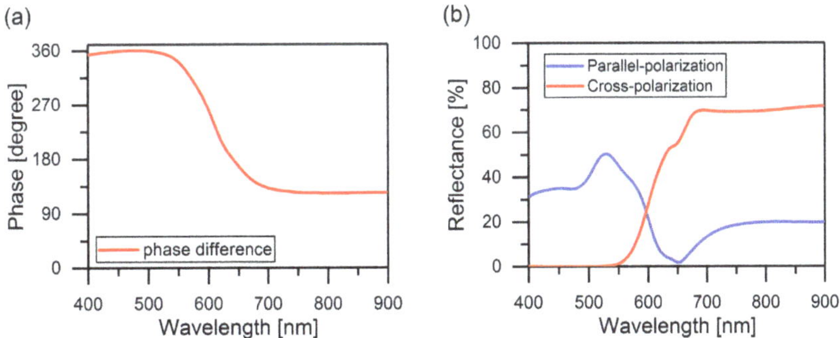

Figure 3. (**a**) Phase difference for reflection from the half-wave plate, with the phase for u-polarized light subtracts the phase of v-polarized light and (**b**) amplitude of circular cross-polarized reflected light, conventional mirror behavior (blue) and co-polarized reflected light, HWP behavior (red) simulated by the software for Multiphysics simulation COMSOL.

For the incidence of right-handed circular light, the electric field of the reflection in circular basis can be written as [38]:

$$E_{\text{ref}} = \frac{1}{2}\left[(r_{\text{u}} - r_{\text{v}})e^{-2i\chi}; (r_{\text{u}} + r_{\text{v}})\right].$$

(1)

Here, χ is the angle between the x and u axis, and r_{u} and r_{v} are the complex reflection coefficients for the u and v axes, respectively. To obtain the unchanged polarization of the reflected light, only the first term of Equation (1) is considered. Therefore, the dimensions of the nanobrick were adjusted to minimize $r_{\text{u}} + r_{\text{v}}$ and maximize $r_{\text{u}} - r_{\text{v}}$. Figure 3b shows the amplitude of the unchanged (co-polarized) and opposite (cross-polarized) handedness of the reflected light. In this case, at a wavelength of 650 nm, a perfect reflection was observed, since the reflection coefficient for co-polarized light was close to 0. Moreover, the HWP effect sharply decreased to the left from the resonance; therefore, the CLC stopband center $\lambda_0 = 750$ nm was chosen to the right from HWP resonance.

According to the temporal coupled-mode theory [39] applied to chiral Tamm state [9], the resonance is described by the reflection amplitude of the total CLC–HWP structure:

$$r_{TP}(\omega) = 1 - \frac{2\gamma_1}{i(\omega_0 - \omega) + (\gamma_1 + \gamma_2)},$$

(2)

where $\gamma_1 = 2kc\exp(-2knL)$, $\gamma_2 = (1-R)kc/2$, $k = \pi\delta\sin 2\chi/\lambda_0$, $\lambda_0/p = n + \sqrt{n^2-1}\cos 2\chi$, $\omega_0 = 2\pi c/\lambda_0$ is resonant cyclic frequency, χ is the angle between the u axis of the HWP–metasurface and the cholesteric director at the interface with the HWP-metasurface, $0° < \chi < 90°$, L is the cholesteric layer thickness, R is the co-handed reflectance of metasurface, for the cholesteric p is the pitch, $n = \sqrt{n_e^2 - n_o^2}/2$ is the average refractive index and $\delta = (n_e^2 - n_o^2)/(n_e^2 + n_o^2)$ is anisotropy.

A novel design combining a CLC and a half-wave plate metasurface is proposed (see Figure 1c,d). Between the CLC and the metasurface, poly(methyl methacrylate) (PMMA) is coated as a layer of 480 nm in thickness to protect the metasurface. Alignment of the CLC is in the direction of the x axis on both the top substrate and bottom protecting layer by a surface rubbing machine. The bonding process combines the superstrate and PMMA–metasurface. Then, CLC was injected into the gap by capillary action, and the thickness of the gap was 1.5 µm. The CLC used has ordinary and extraordinary refractive indices of $n_o = 1.52$ and $n_e = 1.75$, respectively. The center wavelength λ_c of the CLC stopband was 750 nm, as calculated by the equation $\lambda_0 = p\langle n\rangle$, where p is the helical pitch and $\langle n\rangle$ is the average refractive index of the CLC. The optical axis of the CLC lies on the x-y plane, where the orientation depends on the position of the z axis along the helical pitch of the liquid crystal. The equivalent permittivity of the dielectric tensor from CLC can be written as [40]:

$$\varepsilon = \varepsilon_0 \begin{bmatrix} \bar{\varepsilon} + \frac{1}{2}\Delta\varepsilon\cos\left(\frac{4\pi z}{p}\right) & \frac{1}{2}\Delta\varepsilon\sin\left(\frac{4\pi z}{p}\right) & 0 \\ \frac{1}{2}\Delta\varepsilon\cos\left(\frac{4\pi z}{p}\right) & \bar{\varepsilon} + \frac{1}{2}\Delta\varepsilon\cos\left(\frac{4\pi z}{p}\right) & 0 \\ 0 & 0 & \bar{\varepsilon} + \frac{1}{2}\Delta\varepsilon \end{bmatrix}.$$

(3)

Here, $\bar{\varepsilon} = (n_e^2 + n_o^2)/2$, $\Delta\varepsilon = (n_e^2 - n_o^2)$, and z represents the position along the helical axis of the planar CLC. Modeling was conducted by using the finite-element method software COMSOL Multiphysics 4.3b and verified by the Berreman matrix method [41].

The reflectance spectrum of Figure 4 was obtained through the Berreman method simulation. It is in good agreement with Equation (2) for $R = 0.7$. When $\gamma_1(L) = \gamma_2(R)$, $\exp(-2nkL) = 2(1-R)$, the optimal cholesteric thickness is $L = 1.5$ µm.

As for the design shown in Figure 1c, the most important point is that the polarization of the reflected light remains the same as that of the incident light, which provides the phase and polarization matching. As illustrated in Figure 1c,d, the energy of light would be localized between the CLC and the half-wave plate to achieve the resonance condition. Therefore, in Figure 4, a narrow reflection dip is clearly observed within the optical stopband of the CLC.

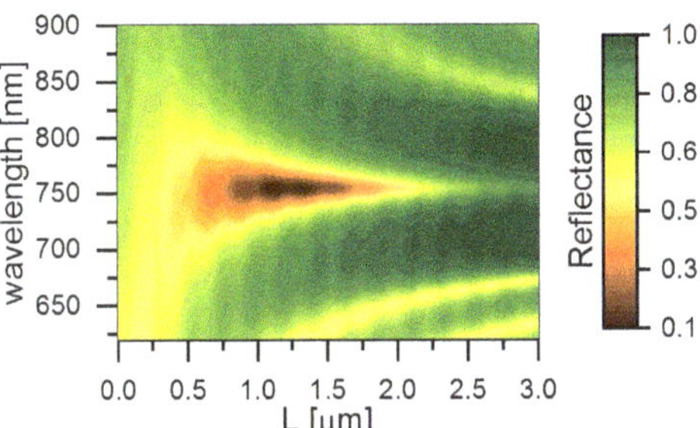

Figure 4. Reflectance spectra of the structure versus the cholesteric layer thickness L simulated by the Berreman method. The incident light is right-handed circular polarized (Figure 1b). Dark dip in the center is the TPP resonance. Reflectance minimum corresponds to the optimal coupling with equal losses through the cholesteric and the metasurface. The resonance bandwidth decreases with increasing L. Small resonances are observed at the edges of cholesteric stopband. Vertical periodic ripples are caused by additional reflection from the upper cholesteric interface; the rippling period equals the half-pitch of cholesteric helix.

As shown in Figure 5, near-field analysis indicates that the strong electric field is localized at the interface [42] between the CLC and the metasurface at the resonance wavelength. In contrast, when the wavelength is nonresonant, the electric-field distribution acts as at an ordinarily-reflecting mirror. The maximum amplitude of the localized electric field at the resonant wavelength is approximately four times larger than that at a nonresonant wavelength. At any point inside the structure the mode has two running wave components with almost equal amplitudes. In Figure 5b, the electric field profile shows spatial ripples due to the interference between the running waves. The ripple period is half-wave. In contrast to a conventional Tamm mode, the interference makes no nodes or antinodes as both running waves are right-handed circularly polarized. In other words, every x-polarized node coincides with a y-polarized antinode and vice versa, which results in a smooth profile [30]. The nontrivial field profile in CLC was thoroughly investigated and illustrated in [43].

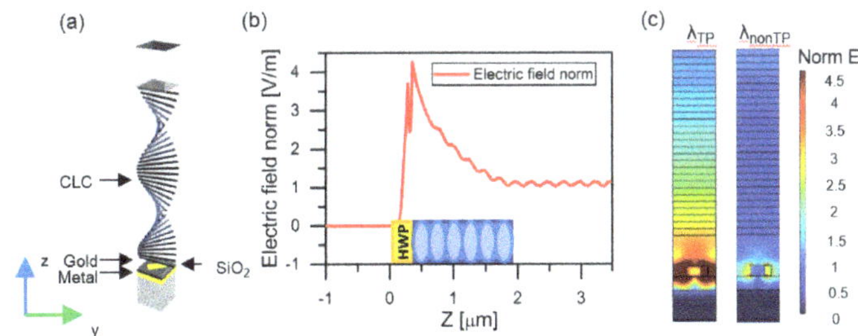

Figure 5. (**a**) Structure of the CLC combined with metasurfaces. (**b**) The maximum of the electric field near CLC–metasurface interface. (**c**) Distribution of normalized electric fields at the resonant and nonresonant wavelengths. Field distribution at the resonant wavelength in *x-y* plane is shown in the Supplementary Materials.

The other advantage of using CLC to generate chiral-selective TPP is that CLC could be easily controlled by external stimuli such as electric field [44] and ambient temperature [45].

By utilizing the temperature dependence of the CLC stopband, the wavelength of resonance can be effectively controlled to achieve a wide-range tunability [46].

As shown in Figure 6, with an increase in temperature from 26 to 29 °C, the resonance dip shifts to a shorter wavelength due to the movement of the CLC stopband. The simulation (dashed lines) and experimental (solid lines) results are in reasonably good agreement. The deviation may result from the thickness of the CLC layer to be different from the setting in the simulated model due to its changes within the measurement area. The imperfection of the rubbing on the PMMA may be an additional reason for the discrepancy between the measured and simulated data. The simulated temperature dependence of the helix pitch was considerably tuned to satisfy Table 1. This restricted our prediction ability for the first experiment. The widening in resonance bandwidth might be due to the extra loss in metal during the nanofabrication process. Table 1 manifests the tunability of the resonance wavelengths of TPP and CLC stopbands with varying temperatures. The difference between the resonant wavelength λ_{TP} and the center wavelength λ_0 is presumably due to the thin protecting layer. The quality factor obtained by coupled mode theory is $Q = \omega_0/2(\gamma_1 + \gamma_2) \approx 27.2$, which is in good agreement with the experimental data.

Table 1. Tunability of the resonant wavelength of chiral-selective TPP (λ_{TP}), center wavelength of CLC stopband (λ_0), pitch p and Q factor of the TPP with respect to the different temperatures.

	26 °C	27 °C	29 °C
λ_{TP} (nm)	809	767	709
λ_0 (nm)	890	805	665
p (nm)	495.4	468.5	428.1
Q factor	28.5	27.2	26.7

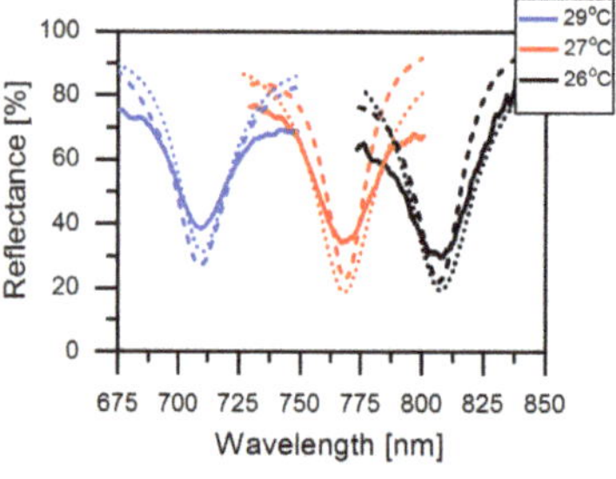

Figure 6. Reflectance spectra of the CLC–metasurface simulated by Berreman method (dotted lines); COMSOL (dashed lines) and experiment results (solid lines) by increasing the temperature from 26 °C to 29 °C.

4. Conclusions

In conclusion, we demonstrated that chiral-selective TPP can be successfully excited at the interface between metasurface of reflective half-wave plate and CLC. This photonic surface state combines properties of both anisotropic metasurface and chiral CLC, providing a wide-ranging orientational tunability. A strong localized electric field at the interface between the CLC and the metasurface was observed. Furthermore, by changing the center wavelength of the stopband of the CLC with different pitches and temperatures, the resonance wavelength of TPP was tuned flexibly. This device can potentially be applied to optical switches and polariton lasers.

Supplementary Materials: The following are available online at https://www.mdpi.com/article/10.3390/ma14112788/s1, Figure S1: Fabrication process of metasurfaces, Figure S2: (a,b) Structure of the bonding devices of CLC and metasurface, which are composed of two plywoods and four precision screws, Figure S3: Schematic of sample fabrication procedure. The plano-convex lens is used as the superstrate to control the thickness of cell gap, and CLC is poured into the gap between the lens and metasurface by capillary action, Figure S4: Experimental setup for reflection measurement. LS: Halogen light source; PH: pinhole; P: polarizer (Thorlabs, WP25M-UB); WP: quarter-wave plate (Edmund, 75 μm in thickness); BS: cube beam-splitter (Thorlabs, BS016); Obj: objective (Olympus, LMPLFLN 50x); S: sample of CLC-metasurface; Detector: spectrometer (Ocean, USB2000+), Figure S5: (a) Measured reflectance for the co-(black line) and cross-polarized (red line) light at normal incidence. Optical image of metasurfaces with (b) cross-polarized measurement and (c) co-polarized measurement, Figure S6: The experimental and simulated reflectance spectra of the structure, Figure S7: Ex (left) and Ey (right) components of the electric field in x-y plane.

Author Contributions: M.-Y.L. and K.-P.C. initiated the work. M.-Y.L. and W.-H.X. performed the optical measurement and characterization. M.-Y.L., J.-H.Y. and C.-R.L. fabricated the sample. M.-Y.L., R.G.B. and I.V.T. performed the numerical simulation and modeling. M.-Y.L., I.V.T., W.L. and K.-P.C. analyzed the experiment data. M.-Y.L., R.G.B., I.V.T., W.L. and K.-P.C. wrote the manuscript. All the authors discussed the results, commented on the manuscript. All authors have read and agreed to the published version of the manuscript.

Funding: This work was supported by the Higher Education Sprout Project of the National Yang Ming Chiao Tung University and Ministry of Education and the Ministry of Science and Technology (MOST-107-2221-E-009-046-MY3; 108-2923-E-009-003-MY3; 109-2224-E-009-002; 109-2628-E-009-007-MY3). The research was also funded by Russian Foundation for Basic Research projects No. 19-52-52006.

Institutional Review Board Statement: Not applicable.

Informed Consent Statement: Not applicable.

Data Availability Statement: The data presented in this study are available upon reasonable request from the corresponding author.

Acknowledgments: Thanks to Taiwan Semiconductor Research Institute (TSRI) and Center for Micro/Nano Science and Technology (CMNST) for the fabrication facility. The fabrication of the metasurfaces were also assisted by Yu-Lun Kuo, Mao-Guo Sun, and Chi-Yin Yang. We especially appreciate the support from Shie-Chang Jeng for supplying the CLC.

Conflicts of Interest: The authors declare no conflict of interest.

References

1. Wu, C.; Arju, N.; Kelp, G.; Fan, J.A.; Dominguez, J.; Gonzales, E.; Tutuc, E.; Brener, I.; Shvets, G. Spectrally selective chiral silicon metasurfaces based on infrared Fano resonances. *Nat. Commun.* **2014**, *5*. [CrossRef] [PubMed]
2. Tang, B.; Li, Z.; Palacios, E.; Liu, Z.; Butun, S.; Aydin, K. Chiral-Selective Plasmonic Metasurface Absorbers Operating at Visible Frequencies. *IEEE Photonics Technol. Lett.* **2017**, *29*, 295–298. [CrossRef]
3. Fan, J.; Lei, T.; Yuan, X. Tunable and Reconfigurable Dual-Band Chiral Metamirror. *IEEE Photonics J.* **2020**, *12*, 1–8. [CrossRef]
4. Mao, L.; Liu, K.; Zhang, S.; Cao, T. Extrinsically 2D-Chiral Metamirror in Near-Infrared Region. *ACS Photonics* **2019**, *7*, 375–383. [CrossRef]
5. Lodahl, P.; Mahmoodian, S.; Stobbe, S.; Rauschenbeutel, A.; Schneeweiss, P.; Volz, J.; Pichler, H.; Zoller, P. Chiral quantum optics. *Nature* **2017**, *541*, 473–480. [CrossRef] [PubMed]
6. Mueller, J.P.B.; Leosson, K.; Capasso, F. Ultracompact metasurface in-line polarimeter. *Optica* **2016**, *3*, 42. [CrossRef]
7. Ma, Q.; Xu, S.Y.; Chan, C.K.; Zhang, C.L.; Chang, G.; Lin, Y.; Xie, W.; Palacios, T.; Lin, H.; Jia, S.; et al. Direct optical detection of Weyl fermion chirality in a topological semimetal. *Nat. Phys.* **2017**, *13*, 842–847. [CrossRef]
8. Jeong, S.M.; Sonoyama, K.; Takanishi, Y.; Ishikawa, K.; Takezoe, H.; Nishimura, S.; Suzaki, G.; Song, M.H. Optical cavity with a double-layered cholesteric liquid crystal mirror and its prospective application to solid state laser. *Appl. Phys. Lett.* **2006**, *89*, 241116. [CrossRef]
9. Timofeev, I.V.; Pankin, P.S.; Vetrov, S.Y.; Arkhipkin, V.G.; Lee, W.; Zyryanov, V.Y. Chiral Optical Tamm States: Temporal Coupled-Mode Theory. *Crystals* **2017**, *7*, 113. [CrossRef]
10. Kaliteevski, M.; Iorsh, I.; Brand, S.; Abram, R.A.; Chamberlain, J.M.; Kavokin, A.V.; Shelykh, I.A. Tamm plasmon-polaritons: Possible electromagnetic states at the interface of a metal and a dielectric Bragg mirror. *Phys. Rev. B* **2007**, *76*, 165415. [CrossRef]
11. Bikbaev, R.; Vetrov, S.; Timofeev, I. Epsilon-Near-Zero Absorber by Tamm Plasmon Polariton. *Photonics* **2019**, *6*, 28. [CrossRef]
12. Jeng, S.C. Applications of Tamm plasmon-liquid crystal devices. *Liq. Cryst.* **2020**. [CrossRef]
13. Zhang, W.L.; Wang, F.; Rao, Y.J.; Jiang, Y. Novel sensing concept based on optical Tamm plasmon. *Opt. Express* **2014**, *22*, 14524. [CrossRef] [PubMed]
14. Qin, L.; Wu, S.; Zhang, C.; Li, X. Narrowband and Full-Angle Refractive Index Sensor Based on a Planar Multilayer Structure. *IEEE Sens. J.* **2019**, *19*, 2924–2930. [CrossRef]

15. Symonds, C.; Lheureux, G.; Hugonin, J.P.; Greffet, J.J.; Laverdant, J.; Brucoli, G.; Lemaitre, A.; Senellart, P.; Bellessa, J. Confined Tamm Plasmon Lasers. *Nano Lett.* **2013**, *13*, 3179–3184. [CrossRef] [PubMed]
16. Lheureux, G.; Azzini, S.; Symonds, C.; Senellart, P.; Lemaître, A.; Sauvan, C.; Hugonin, J.P.; Greffet, J.J.; Bellessa, J. Polarization-Controlled Confined Tamm Plasmon Lasers. *ACS Photonics* **2015**, *2*, 842–848. [CrossRef]
17. Lheureux, G.; Monavarian, M.; Anderson, R.; Decrescent, R.A.; Bellessa, J.; Symonds, C.; Schuller, J.A.; Speck, J.S.; Nakamura, S.; DenBaars, S.P. Tamm plasmons in metal/nanoporous GaN distributed Bragg reflector cavities for active and passive optoelectronics. *Opt. Express* **2020**, *28*, 17934. [CrossRef] [PubMed]
18. Zhang, X.L.; Song, J.F.; Li, X.B.; Feng, J.; Sun, H.B. Optical Tamm states enhanced broad-band absorption of organic solar cells. *Appl. Phys. Lett.* **2012**, *101*, 243901. [CrossRef]
19. Bikbaev, R.G.; Vetrov, S.Y.; Timofeev, I.V.; Shabanov, V.F. Photosensitivity and reflectivity of the active layer in a Tamm-plasmon-polariton-based organic solar cell. *Appl. Opt.* **2021**, *60*, 3338. [CrossRef]
20. Timofeev, I.V.; Arkhipkin, V.G.; Vetrov, S.Y.; Zyryanov, V.Y.; Lee, W. Enhanced light absorption with a cholesteric liquid crystal layer. *Opt. Mater. Express* **2013**, *3*, 496. [CrossRef]
21. Johnson, P.; Christy, R.W. Optical constants of the noble metals. *Phys. Rev. B* **1972**, *6*, 4370–4379. [CrossRef]
22. Sultanova, N.; Kasarova, S.; Nikolov, I. Dispersion Properties of Optical Polymers. *Acta Phys. Pol. A* **2009**, *116*, 585–587. [CrossRef]
23. Vetrov, S.Y.; Pyatnov, M.V.; Timofeev, I.V. Spectral and polarization properties of a 'cholesteric liquid crystal—Phase plate—Metal' structure. *J. Opt.* **2015**, *18*, 015103. [CrossRef]
24. Hao, J.; Yuan, Y.; Ran, L.; Jiang, T.; Kong, J.A.; Chan, C.T.; Zhou, L. Manipulating Electromagnetic Wave Polarizations by Anisotropic Metamaterials. *Phys. Rev. Lett.* **2007**, *99*. [CrossRef]
25. Xiao, S.; Mühlenbernd, H.; Li, G.; Kenney, M.; Liu, F.; Zentgraf, T.; Zhang, S.; Li, J. Helicity-Preserving Omnidirectional Plasmonic Mirror. *Adv. Opt. Mater.* **2016**, *4*, 654–658. [CrossRef]
26. Pors, A.; Nielsen, M.G.; Bozhevolnyi, S.I. Broadband plasmonic half-wave plates in reflection. *Opt. Lett.* **2013**, *38*, 513. [CrossRef]
27. Damgaard-Carstensen, C.; Ding, F.; Meng, C.; Bozhevolnyi, S.I. Demonstration of $>2\pi$ reflection phase range in optical metasurfaces based on detuned gap-surface plasmon resonators. *Sci. Rep.* **2020**, *10*. [CrossRef] [PubMed]
28. Ding, F.; Chen, Y.; Bozhevolnyi, S.I. Gap-surface plasmon metasurfaces for linear-polarization conversion, focusing, and beam splitting. *Photonics Res.* **2020**, *8*, 707. [CrossRef]
29. Deshpande, R.A.; Ding, F.; Bozhevolnyi, S. Dual-Band Metasurfaces Using Multiple Gap-Surface Plasmon Resonances. *ACS Appl. Mater. Interfaces* **2019**, *12*, 1250–1256. [CrossRef]
30. Avdeeva, A.Y.; Vetrov, S.Y.; Bikbaev, R.G.; Pyatnov, M.V.; Rudakova, N.V.; Timofeev, I.V. Chiral Optical Tamm States at the Interface between a Dye-Doped Cholesteric Liquid Crystal and an Anisotropic Mirror. *Materials* **2020**, *13*, 3255. [CrossRef]
31. Timofeev, I.; Vetrov, S.Y. Chiral optical Tamm states at the boundary of the medium with helical symmetry of the dielectric tensor. *JETP Lett.* **2016**, *104*, 380–383. [CrossRef]
32. Rudakova, N.V.; Timofeev, I.V.; Bikbaev, R.G.; Pyatnov, M.V.; Vetrov, S.Y.; Lee, W. Chiral Optical Tamm States at the Interface between an All-Dielectric Polarization-Preserving Anisotropic Mirror and a Cholesteric Liquid Crystal. *Crystals* **2019**, *9*, 502. [CrossRef]
33. Zhang, L.Y.; Yin, X.; Yang, J.; Li, A.; Xu, G.K. Multilevel structural defects-induced elastic wave tunability and localization of a tensegrity metamaterial. *Compos. Sci. Technol.* **2021**, *207*, 108740. [CrossRef]
34. Liu, Y.; Cheng, J.; Yang, H.; Xu, G.K. Rotational constraint contributes to collective cell durotaxis. *Appl. Phys. Lett.* **2020**, *117*, 213702. [CrossRef]
35. Zheng, G.; Mühlenbernd, H.; Kenney, M.; Li, G.; Zentgraf, T.; Zhang, S. Metasurface holograms reaching 80% efficiency. *Nat. Nanotechnol.* **2015**, *10*, 308–312. [CrossRef] [PubMed]
36. Dong, Y.; Xu, Z.; Li, N.; Tong, J.; Fu, Y.H.; Zhou, Y.; Hu, T.; Zhong, Q.; Bliznetsov, V.; Zhu, S.; et al. Si metasurface half-wave plates demonstrated on a 12-inch CMOS platform. *Nanophotonics* **2019**, *9*, 149–157. [CrossRef]
37. Dorrah, A.H.; Rubin, N.A.; Zaidi, A.; Tamagnone, M.; Capasso, F. Metasurface optics for on-demand polarization transformations along the optical path. *Nat. Photonics* **2021**, *15*, 287–296. [CrossRef]
38. Jones, R.C. A New Calculus for the Treatment of Optical Systems IV. *J. Opt. Soc. Am.* **1942**, *32*, 486. [CrossRef]
39. Joannopoulos, J.D.; Johnson, S.G.; Winn, J.N.; Meade, R.D. *Photonic Crystals: Molding the Flow of Light*, 2nd ed.; Princeton Univ. Press: Princeton, NJ, USA, 2008; p. 304.
40. Kallos, E.; Yannopapas, V.; Photinos, D.J. Enhanced light absorption using optical diodes based on cholesteric liquid crystals. *Opt. Mater. Express* **2012**, *2*, 1449. [CrossRef]
41. Berreman, D.W. Optics in Stratified and Anisotropic Media: 4×4-Matrix Formulation. *J. Opt. Soc. Am.* **1972**, *62*, 502. [CrossRef]
42. Chang, C.Y.; Chen, Y.H.; Tsai, Y.L.; Kuo, H.C.; Chen, K.P. Tunability and Optimization of Coupling Efficiency in Tamm Plasmon Modes. *IEEE J. Sel. Top. Quantum Electron.* **2015**, *21*, 262–267. [CrossRef]
43. Fang, X.; MacDonald, K.F.; Plum, E.; Zheludev, N.I. Coherent control of light-matter interactions in polarization standing waves. *Sci. Rep.* **2016**, *6*. [CrossRef]

44. Huang, Y. Polarization independent two-way variable optical attenuator based on polymer-stabilized cholesteric liquid crystal. *Opt. Express* **2010**, *18*, 10289. [CrossRef] [PubMed]
45. Hsiao, Y.C.; Wang, H.T.; Lee, W. Thermodielectric generation of defect modes in a photonic liquid crystal. *Opt. Express* **2014**, *22*, 3593. [CrossRef] [PubMed]
46. Hsiao, Y.C.; Yang, Z.H.; Shen, D.; Lee, W. Red, Green, and Blue Reflections Enabled in an Electrically Tunable Helical Superstructure. *Adv. Opt. Mater.* **2018**, *6*, 1701128. [CrossRef]

MDPI

St. Alban-Anlage 66

4052 Basel

Switzerland

Tel. +41 61 683 77 34

Fax +41 61 302 89 18

www.mdpi.com

Materials Editorial Office

E-mail: materials@mdpi.com

www.mdpi.com/journal/materials